Isoform p53 Protein's Major Role in the Pathophysiology of Malignant Hematologic Diseases

Isoform p53 Protein's Major Role in the Pathophysiology of Malignant Hematologic Diseases

By

Aurelian Udristioiu, Nica-Badea Delia
and Ioan-Ovidiu Gheorghe

Cambridge
Scholars
Publishing

Isoform p53 Protein's Major Role in the Pathophysiology
of Malignant Hematologic Diseases

By Aurelian Udristioiu, Nica-Badea Delia and Ioan-Ovidiu Gheorghe

This book first published 2024

Cambridge Scholars Publishing

Lady Stephenson Library, Newcastle upon Tyne, NE6 2PA, UK

British Library Cataloguing in Publication Data
A catalogue record for this book is available from the British Library

ISBN (10): 1-5275-7243-9
ISBN (13): 978-1-5275-7243-0

CONTENTS

FOREWORD

The book "**Isoform p53 Protein's Major Role in the Pathophysiology of Malignant Hematologic Diseases**" is concerned with the identification of isoform p53 protein in malignant hematologic diseases and its impact on the clinical management of patients.

This research book presents an extensive study of malignant B cells, with presented isoform p53 protein, in Chronic Lymphocytic Leukemia (CLL). Cases of CLL were analyzed with the ELISA system to evaluate the relationship between isoform p53 protein and disease status, and the possible impact of this protein on responses to treatment and the survival of patients.

The book also provides important benchmarks in the field, studying the hematological and metabolic problems of the human body. The book details advances made over the past decade in cancer metabolism research that have increased the understanding of aerobic and anaerobic glycolysis and other metabolic changes associated with cell growth and proliferation by altering cell signaling and blocking their differentiation in malignancies.

The book, which has original content, deals not only with theoretical aspects, but also with practical aspects, highlighting the need for new practices in the biological sciences. It takes into account the progress in the pathophysiology and biochemistry of oncology over the years.

The new book, published by Cambridge Scholars Publishing, Great Britain, may be of interest to doctors of all specialties, medical students, as well as staff working in clinical laboratories, from different countries. It presents a large volume of data, experiments and laboratory research in a systematic and logical display.

Aurelian Udristioiu

ACKNOWLEDGEMENT

Dr Cristiana Tanase, Titu Maiorescu University of Bucharest, Faculty of Medicine, Professor Molecular Biology, Clinical Biochemistry and Research Management, PhD theses coordinator, MD, PhD.

CHAPTER ONE

INTRODUCTION TO THE FUNCTION OF THE p53 GENE AS A TUMOR SUPPRESSOR GENE

The p53 gene is a tumor suppressor gene and its activity stops the formation of tumors. If a person inherits only one functional copy of the p53 gene from their parents, they are predisposed to cancer and usually develop several independent tumors in a variety of tissues in early adulthood. Mutations in the p53 gene are found in most tumor types, and so contribute to the complex network of molecular events leading to tumor formation.

The International Cancer Genome Consortium has established that the p53 gene is the most frequently mutated (>50%) in human cancer, indicating that the p53 gene plays a crucial role in preventing cancer formation. The p53 gene encodes proteins that bind to DNA and regulate gene expression to prevent mutations in the genome.

Warren Maltzman, of the Waksman Institute of Rutgers University, first demonstrated that p53 was responsive to DNA damage in the form of ultraviolet radiation. In a series of publications in 1991–1992, Michael Kastan, Johns Hopkins University, reported that TP53 was a critical part of a signal transduction pathway that helped cells respond to DNA damage. In 1993, p53 was voted molecule of the year by Science magazine, USA.

The p53 gene has been mapped to chromosome 17. In the cell, its product, the p53 protein, binds DNA, which in turn stimulates another gene, CDKN, to produce a protein called p21 that interacts with a cell division-stimulating protein (cdk2).

When the p21 protein is complex with the cdk2 protein, the cell cannot pass through to the next stage of cell division. Therefore, the isoform p53 protein as a result of the p53 mutated gene can no longer effectively bind

DNA, and as a consequence, the p21 protein is not made available to act as the 'stop signal' for cell division.

Although the number of genes activated by the p53 gene is rather large, the outcome of p53 protein activation is either cell cycle arrest in G1 (by p21) or apoptosis (by BAX, PUMA or NOXA). The cell growth arrest activity of p53 protein allows the activation of the DNA repair system of the cell.

The amount of information that exists on all aspects of p53's normal function and mutant expression in human cancers is now vast, reflecting its key role in the pathogenesis of human cancers. Mutant p53 genes are just one component of a network of events that culminate in tumor formation.

p53 gene alterations have been found in: 20-30% of chronic myeloid leukemia (CML) blast phases (or blast crises), often associated with 17q chromosome deletion; in 15% of Burkitt lymphomas; in 40% of aggressive chronic lymphocytic leukemia (CLL) cases, with transformation into Richter's syndrome; in 5-10% of multiple myelomas; in 5% of myelodysplastic syndrome (MDS) cases; in 60-80% of Hodgkin disease cases; and in 30% of high-grade B-cell non-Hodgkin lymphoma cases. p53 gene alterations in hematological malignancies are associated with a poor prognosis.

Inherited changes in the p53 gene greatly increase the risk of developing forms of cancer, as part of a rare cancer syndrome called Li-Fraumeni syndrome (LES). Most of these mutations change single protein building block amino acids in the p53 protein, which reduces or eliminates the protein's tumor suppressor function. Because the altered protein is less able to regulate cell growth and division, DNA damage can accumulate.

This damage may contribute to the development of a cancerous tumor by allowing cells to grow and divide. Although somatic mutations in the p53 gene are found in many types of cancer, LES appears to be the only cancer syndrome associated with inherited mutations in this gene.

This condition greatly increases the risk of developing several types of cancer, particularly in children and young adults. At least 140 different mutations of the p53 gene have been identified in individuals with LES.

Many of the mutations associated with LES change single amino acids in the part of the p53 protein that binds to DNA. Other mutations delete small amounts of DNA from the gene. Mutations in the p53 gene lead to a version of the p53 protein that cannot regulate cell growth and division effectively. Specifically, the altered protein is unable to trigger apoptosis in cells with mutated or damaged DNA. As a result, DNA damage can accumulate in cells. Such cells may continue to divide in an uncontrolled way, leading to the growth of tumors [1].

CHAPTER TWO

NORMAL FUNCTION OF THE P53 GENE

1.1. p53 protein discovery

The p53 protein (p53) was first identified in 1979 by Lionel Crawford, David P. Lane, Arnold Levine, and Lloyd Old, working at the Imperial Cancer Research Fund (UK), Princeton University, UMDNJ (Cancer Institute of New Jersey), and the Memorial Sloan-Kettering Cancer Center, respectively. Independently, it was also identified by Michel Kress, Pierre May, José A. Melero and Varda Rotter. It had been hypothesized to exist beforehand, as the target of the SV40 virus, a strain that induced the development of tumors. The human p53 gene was cloned in 1984.

The name p53 was given in 1979 to describe the apparent molecular mass; SDS-PAGE analysis indicated that it was a 53 kilodalton (k Da) protein. However, the actual mass of the full-length p53 protein (p53α), based on the sum of masses of the amino acid residues, is only 43.7 k Da. In addition to the full-length protein, the human p53 gene encodes at least 15 protein isoforms, ranging in size from 3.5 to 43.7 k Da. All these p53 proteins are called the p53 isoforms**.** In humans, the p53 gene is located on the short arm of chromosome 17 (17p13.11) [2].

Researchers have extensively studied factors known to damage DNA, such as UV light and tobacco smoke, along with the cellular mechanisms of DNA repair.

Cells also naturally accumulate a certain number of mutations with each division. If all mammalian cells were equally susceptible to mutations that could lead to cancer, then cancer risk should increase with the greater number of cells, species life span and greater number of cell divisions **[Scheme 1]**.

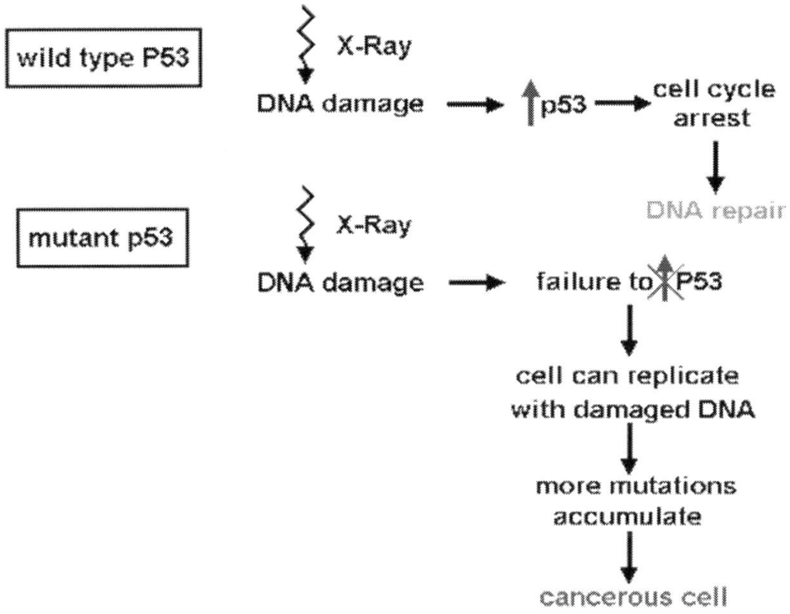

Scheme 1. *Potential path of mutant p53 genes leading to cancerous cells.*

However, researchers found no significant relationships between cancer risk and body size, life span, or basic metabolic rate among the species. For elephants, they estimated that the overall lifetime chance of dying from cancer was less than 5%. The lifetime cancer mortality rate for humans, by contrast, is about 20%. To gain insight into how elephants avoid cancer, researchers looked at elephant genomes, with a focus on the p53 gene.

The team analyzed genomic sequences and found that African elephants have at least 40 p53 alleles. Functional assays comparing elephant and human cells showed that in response to DNA damage, cells from elephants had higher cell death (apoptosis) rates but didn't boost DNA repair mechanisms.

Both are known p53-mediated responses to DNA damage. Cells obtained from people with LES had significantly less apoptosis than the human controls and more than 5 times less than elephants. These results suggest that elephants may have evolved to resist cancer by triggering apoptosis through the p53 protein to efficiently remove mutant cells (Harrison Wein, October 26, 2015. How Elephants Defend Against Cancer, NIH, USA).

The WHO defines a single category of myelodysplastic syndrome (MDS) with biallelic p53 gene inactivation, irrespective of the blast percentage, but excludes single-hit p53 mutant genes in MDS with bone marrow (BM) blasts <20%. Similarly, the recent International Prognostic Scoring System-Molecular acknowledged the poor outcome of the multi-hit p53 mutant genes but excluded the single-hit p53 mutant gene.

The ICC and ELN guidelines emphasize a p53 mutant variant allele frequency (VAF) of >10% regardless of the single- or multi-hit status of the genes for MDS/acute myeloid leukemia (AML) [3] **[Scheme 2]**.

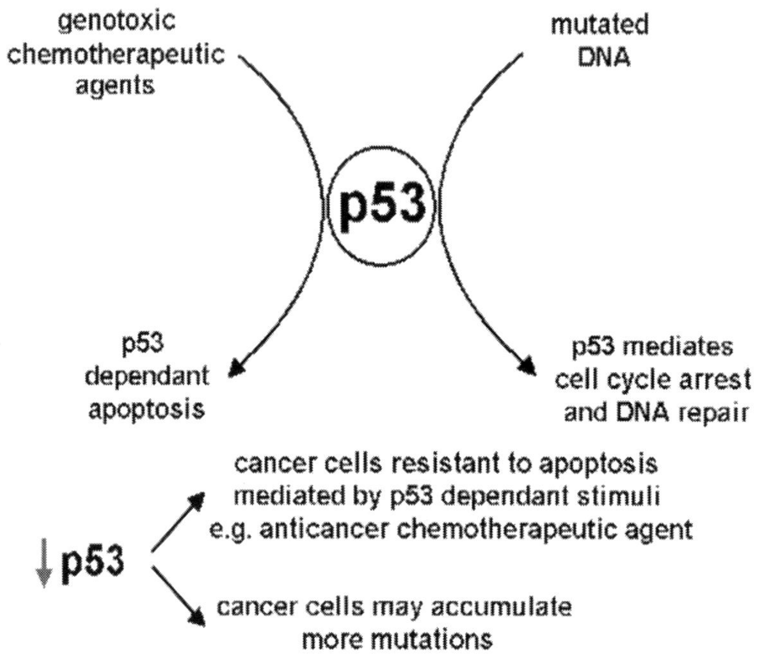

Scheme 2*. A mutation of the p53 gene causing it to lose any of its functions will inevitably lead to carcinogenesis.*

1.2. Regulation of Isoform p53 Protein

As with 95% of human genes, the p53 gene encodes more than one protein. In 2005 several isoforms were discovered, and as of now, 12 human p53 isoforms have been identified (p53α, p53β, p53γ, Δ40p53α, Δ40p53β, Δ40p53γ, Δ133p53α, Δ133p53β, Δ133p53γ, Δ160p53α, Δ160p53β, and Δ160p53γ). Furthermore, p53 isoforms are expressed in a tissue-dependent manner and p53α is never expressed alone.

The full length isoform p53 proteins can be subdivided into different protein domains. Starting from the N-terminus, there are first the amino-terminal transactivation domains (TAD 1 and TAD 2), which are needed to induce a subset of p53 target genes. These domains are followed by the Proline rich domain (PXXP), whereby the motif PXXP is repeated (P is a Proline and X can be any amino acid). This is required, among other things, for p53 mediated apoptosis. Some isoforms lack the Proline rich domain, such as Δ133p53βγ and Δ160p53αβγ, hence some isoforms of p53 do not mediate apoptosis, which emphasizes the diverse roles of the p53 gene.

Next, there is the DNA binding domain (DBD), which enables the proteins to sequence specific bindings. The carboxyl terminal domain completes the protein. It includes the nuclear localization signal (NLS), the nuclear export signal (NES) and the oligo-merization domain (OD) [4].

Isoform p53 proteins often fail to induce MDM2 protein, which causes p53 protein to accumulate at very high levels. Moreover, the isoform p53 proteins themselves can inhibit normal p53 protein levels. The dynamics of p53 proteins, along with its antagonist, MDM2 protein, indicate that the levels of p53, in units of concentration, oscillate as a function of time.

This "dumped" oscillation is both clinically documented and mathematically modelled. Current models can also be useful for the modelling of p53 isoforms and their effects on p53 oscillation, thereby promoting *de novo* tissue-specific pharmacological drug discovery.

The protein kinases that are known to target the transcriptional activation domain of p53 protein can be roughly divided into two groups. The first group of protein kinases belongs to the MAPK family (JNK1-3, ERK1-2, and p38 MAPK), which is known to respond to several types of stress, such as membrane damage, oxidative stress, osmotic shock, heat shock, etc.

The second group of protein kinases (ATR, ATM, CHK1, CHK2, DNA-PK, CAK, and P53RK) is implicated in the genome integrity checkpoint, a molecular cascade that detects and responds to several forms of DNA damage caused by genotoxic stress. Oncogenes also stimulate p53 protein activation, mediated by the protein p14ARF. The critical event leading to the activation of p53 protein is the phosphorylation of its N-terminal domain.

The N-terminal transcriptional activation domain contains a large number of phosphorylation sites and can be considered as the primary target for protein kinases transducing stress signals [5].

Phosphorylation of the N-terminal end of the p53 protein by the above-mentioned protein kinases disrupts MDM2-binding. Other proteins, such as Pin1, are then recruited to p53 and induce a conformational change in p53, which prevents MDM2-binding even more. Phosphorylation also allows for the binding of transcriptional coactivators, like p300 and PCAF, which then acetylate the carboxy-terminal end of p53 protein, exposing the DNA binding domain of p53 protein, allowing it to activate or repress specific genes. Deacetylase enzymes, such as Sirt1 and Sirt7, can deacetylate p53 protein, leading to the inhibition of apoptosis. Some oncogenes can also stimulate the transcription of proteins that bind to MDM2 and inhibit its activity.

1.3. Function of p53 protein

Activated p53 protein binds to DNA and activates the expression of several genes, including microRNA miR-34a and WAF1/CIP1, thereby encoding p21 protein and hundreds of other downstream gene binds to the G1-S/CDK (CDK4, CDK6, CDK2, and CDK1) complexes (molecules important for the G1/S transition in the cell cycle), inhibiting their activity. When p21 protein (WAF1) is complexed with CDK2, the cell cannot continue to the next stage of cell division. An isoform p53 will no longer bind to DNA in an effective way, and, as a consequence, the p21 protein will not be available to act as the "stop signal" for cell division.

Studies of human embryonic stem cells (h ESCs) commonly describe the nonfunctional p53-p21 axis of the G1/S checkpoint pathway with subsequent relevance for cell cycle regulation and the DNA damage response (DDR).

Mathematical models also indicate that the p53 concentration oscillates much faster once teratogens, such as double-stranded breaks (DSB) or UV radiation, are introduced to the system. This supports and models the current understanding of p53 protein dynamics, whereby DNA damage induces p53 protein activation (see p53 regulation for more information). Current models can also be useful for modelling the mutations in p53 isoforms and their effects on p53 oscillation, thereby promoting de novo tissue-specific pharmacological drug discovery [6].

CHAPTER THREE

STRUCTURE OF THE P53 GENE

The p53 gene is located in the 17p13.1 chromosome and has: molecular location in chromosome 17; base pairs 7,668,402 to 7,687,550; and homo sapiens annotation release 107, GRCh38.p2. People have 2 alleles (copies) of the p53 gene. One is inherited from each parent, and both are crucial to prevent cancer. The p53 gene is mutated in most human cancers. Having only one functional allele causes LES, which is characterized by a more than 90% lifetime risk of cancer **[Scheme 3]**.

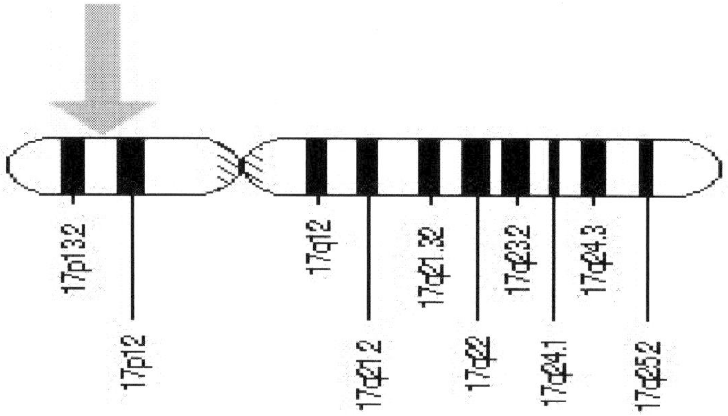

Scheme 3. *The TP53 gene is located on the short (p) arm of chromosome 17 at position 13.1. More precisely, the TP53 gene is located from base pair 7,668,402 to base pair 7,687,550 on chromosome 17.*

The p53 chip scans exons 2-11, containing more than 98% of all mutations described so far in the gene. This system has been designed to allow the detection of the most common mutations and all identified polymorphisms in the p53 coding sequence. It has been found that this system allows for the sequencing of an average of 97.5% of the arrayed p53 genes.

Altogether, 1329 bp of the sequence is analyzed from sense and anti-sense strands. The p53 chip detection limit for known alleles has been identified at 5%. The actual limit could sometimes be even less than 5%, but in real life, the possible alleles are mostly unknown and reliable controls and comparisons with results obtained with standard methods can be technically difficult because of their error rates.

The p53 chip has been tested in cooperation with partners from the IARC. In solid tumors, the p53 gene is mutated or deleted in about 50% of all tumors, but in leukemia, aberrations of the p53 gene are infrequent, ranging from 5-10% at diagnosis (Peller & Rotter, 2003). Mutations in the p53 gene occur mostly in the DNA binding domain (exons 5-8), as missense or nonsense mutations, deletions or inserts.

The p53 protein functions as a tetramer and the isoform p53 protein and the wildtype p53 protein form a tetrameric complex, while the mutant p53 acts as a dominant negative inhibitor of the wildtype p53 (loss-of-function) (Vousden & Lu, 2002). Patients with p53 gene mutations are not recognized at diagnosis by routine diagnostic screening with FISH.

Like patients with 17p del, patients with mutant p53 genes show poor survival rates and resistance to chemotherapy. Among patients with chemotherapy-refractory disease, about 40-50% have lost one 17p allele or have a mutated p53 gene, leading to reduced overall survival rates. More than 95% of the identified p53 mutations are found in exons 5-8, the DNA binding domain (Rossi et al., 2009).

In de novo acute myeloid leukemia (AML), the loss of the short arm of chromosome 17 is reported in 5-10% of cases (Stirewalt et al., 2001; Seifert et al., 2009). However, the deletion is mostly accompanied by complex aberrant karyotypes and is rare in other subgroups.

As in chronic lymphocytic leukemia (CLL), 17p del in one allele is most often associated with TP53 point mutations in the remaining allele, with adverse overall survival rates and low resistance to conventional chemotherapy. Haferlach et al. (2008) and Rűcker et al. (2012) have reported p53 mutation rates as high as 60-70%, respectively, in AML with complex karyotypes, most often associated with deletions of chromosomes 5, 7 and 17p.

TP53 mutations in AML in any of the risk groups have been seen to represent a specific subset with the most un-favorable outcomes, displaying a very short OS (Grossmann et al., 2012).

A mutation of the p53 gene causing it to lose any of its functions will inevitably lead to carcinogenesis, which will allow the cell to grow indefinitely without any regulation. Another important factor in the process of carcinogenesis is the balance between the pro-apoptotic and anti-apoptotic members of the Bcl-2 family.

In a tumor cell, a mutation in the Bcl-2 gene results in increased expression and will suppress the normal function of the pro-apoptotic proteins BAX and BAK, leading to malignancy.

On the other hand, a mutation in the BAX or BAK genes can cause a down-regulation of expression, causing the cell to lose its ability to regulate apoptosis, once again leading to cancer cells. The inhibitor of apoptosis (IAP) family genes, which encodes negative regulatory proteins, can prevent apoptotic cell death. In a normal cell, the p53 protein binds to DNA, stimulating another gene to produce a protein called p21, which interacts with a cell division-stimulating protein (cdk2) [7].

The disruption of surviving microtubule interactions results in the loss of the anti-apoptotic function of the survivin protein and increased caspase-3 activity, a mechanism involved in cell death during mitosis. It has been suggested that the survivin protein may counteract the default induction of apoptosis in the G2/M phase **[Scheme 4]**.

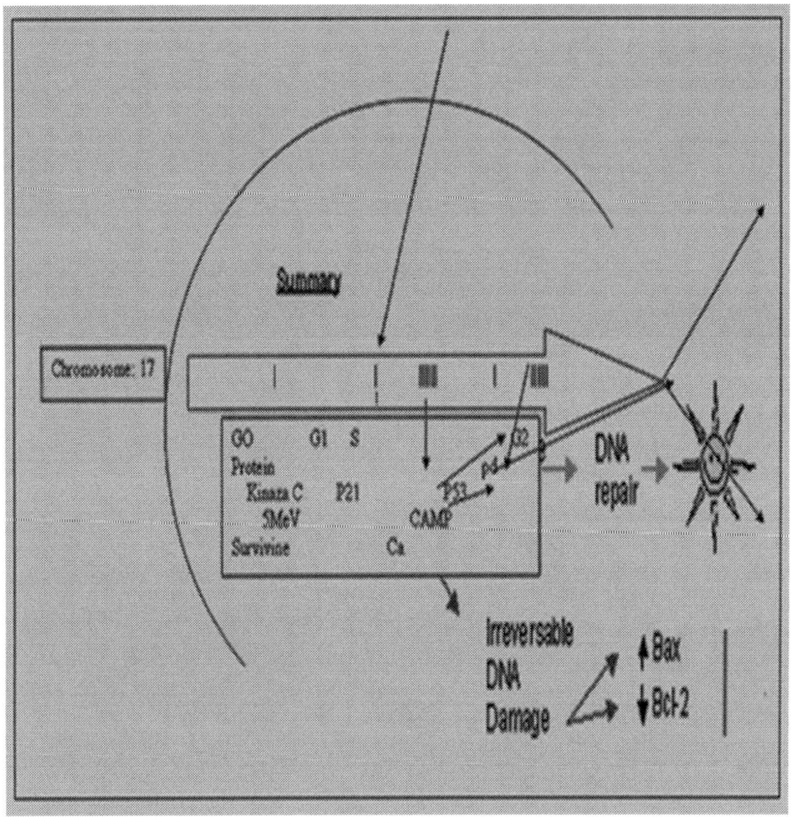

Scheme 4. *In the normal cell, the p53 protein binds to DNA, stimulating another gene to produce a protein called p21, which interacts with a cell division-stimulating protein (cdk2). [Udristioiu A., Florescu C., Popescu MA., Cojocaru M. High Concentration of anaerobic ATP implicated in aborted apoptosis from CLL. Lab Medicine 2010; 41: 203-08].*

Activation of miR-34a and miR-145 can maintain pluripotency by inhibiting stem cell markers such as the Lin28a, Oct4, Klf4, and Sox2 genes. On the other hand, it has been observed that the homozygous deletion of the p53 gene can lead to a stemless phenotype in pancreatic acinar cells showing elevated expression of cancer stem cell (CSC) markers and stem cell regulators such as c-Myc gene, SOX9, and Klf4, along with several other genes. Therefore, the loss of the p53 gene can also result in the development of stem cell-like characteristics that enhance tumor growth.

There are many p53 gene-induced mRNAs and encoded proteins involved in a variety of potentially anti-neoplastic processes, including inhibition angiogenesis, DNA repair, cell cycle checkpoints, senescence and apoptosis (Vogelstein et al., 2000; Riley et al., 2008; Brady and Attardi, 2010).

The activation of consecutive p53 gene responses, separated by a period of recovery, leads to the selective attenuation of a subset of p53-regulated mRNAs corresponding to those targeted by one or more of the p53-responsive miRNAs. Our results indicate that the long-term expression of p53-responsive miRNAs leads to an excess of miRNAs during the second response, and this likely prevents the induction of MDM2, BTG2 and CDKN1A mRNA and/or protein.

These observations are likely to have important implications for daily cancer therapies that activate p53 protein in normal tissues and/or tumor cells. The p53 protein regulates the expression of a subset of miRNAs, noncoding RNAs that contribute to post-transcriptional gene silencing, linking the p53 protein response to post-transcriptional regulation of gene expression in a separate way (Suzuki et al., 2009; He et al., 2007b; He et al., 2007c). Several recently identified miRNAs strongly induced in HT29-tsp53 cells are: miR-143-3p, miR-145-5p, miR-139-5p and miR-34a-5p (Cabrita et al., 2016).

These mRNAs were predominantly short-lived, permitting rapid recovery from p53 activation and rapid cell cycle re-entry. Here, we similarly detected the increased expression of p53-responsivness at miRNAs containing the miR-34, miR-139, miR-192 and miR-194 hairpins, and this was once again readily reversed at the restrictive temperature. The reversible expression of these mRNAs was not entirely surprising for two reasons. First, the p53-transcriptional mRNA response is reversed within 1 h at the restrictive temperature.

Ribonucleic acid miRNAs are similarly transcribed by RNA polymerase II, so their p53-dependent transcription is likely done through the same fundamental mechanism. It is suggested that this favors functional miRNA-mRNA interactions that selectively attenuate targeted gene expression. Although it remains to be tested, this type of delayed feedback may be a common feature of other transcriptional responses. One could hypothesize that interacting miRNAs and mRNAs could be co-expressed in progenitor cell populations [8] **[Scheme 5]**.

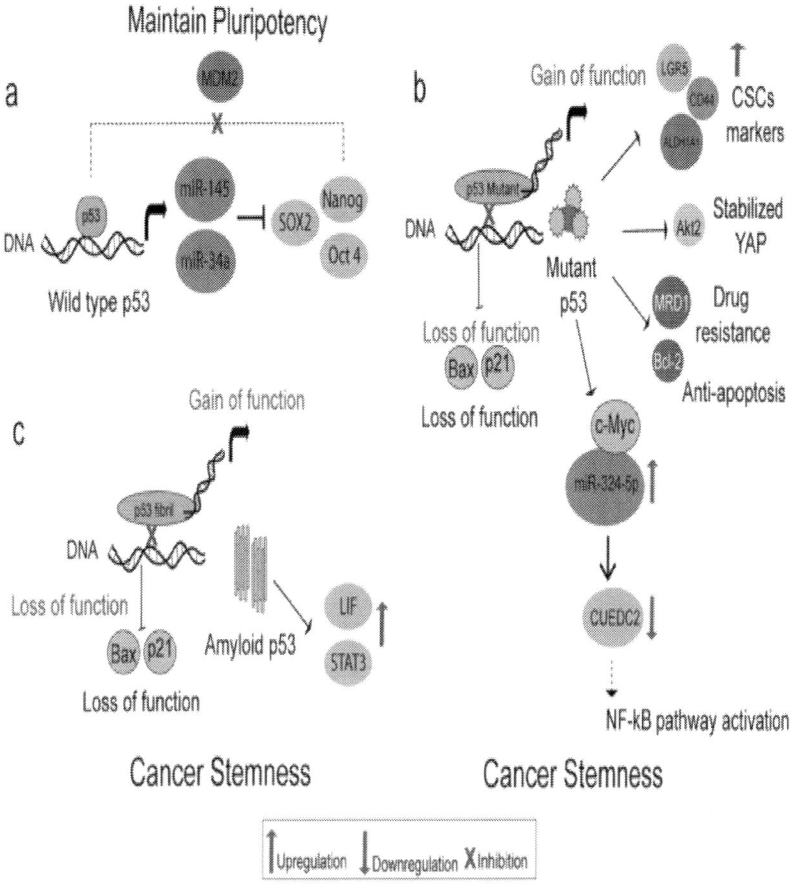

Scheme 5. *Long-term expression of miRNAs in differentiating cell populations could contribute to the long-term silencing of specific mRNAs, preventing ectopic mRNA expression.*

A temperature-sensitive variant of p53 protein that permits rapid reversible activation of the p53 protein response was used to compare the induction and decay of mRNAs and miRNAs throughout two separately defined p53 responses. This resulted in evidence that p53-induced miRNAs provide delayed post-transcriptional feedback that inhibits the expression of

MDM2, BTG2 and p21WAF1 following an initial period of recovery. While the work presented here lacks functional assays, our analysis coupled with functional analysis from other laboratories builds a viable and novel working model for delayed post-transcriptional feedback that will be tested in future studies.

Based on all the above instances, it is evident that isoform p53 proteins and amyloid p53 proteins play a vital role in tumor initiation, progression, and metastasis. Several of the p53 isoforms are known to form p53 amyloids and display loss of function (LOF) and gain of function (GOF) phenotypes. The entire road of cancer progression is controlled by p53 protein via several interacting signaling molecules and pathways. Therefore, any therapeutic approach leading towards p53 protein reactivation could be of great importance to the field of precision medicine.

Specific germline mutations in hereditary breast and ovarian cancer susceptibility (HBC/HBOC) genes - the BRCA1, BRCA2 and PALB2 genes - have been shown to recur in French Canadians from Quebec, Canada, and this has been attributed to common ancestors. Germline p53 mutation carriers are known in Li-Fraumeni syndrome families, and feature the onset of breast cancer at a young age.

Reportedly, a rare p53 mutation was carried in French Canadian HBC families, though none recurred, possibly due to the limited number of cancer families investigated.

Information on p53 germline mutations found in French Canadian cancer families and provided by hereditary cancer clinics led to the investigation of 37 new BRCA1 and BRCA2 mutation-negative HBC/HBOC families for the p53 mutations, and the assessment of the frequency of p53 mutations in 1,235 French Canadian breast cancer cases not selected for family history of cancer.

p53 gene mutation-positive pedigrees from French Canadian cancer families were provided by local hereditary cancer clinics. Bi-directional Sanger sequencing of all the protein-encoding exons of p53 was performed using peripheral blood lymphocyte DNA from breast/ovarian cancer probands from 37 HBC/HBOC Canadian families. Targeted bi-directional Sanger sequencing assays of regions containing the identified p53 gene mutations was performed on 1,235 French Canadian breast cancer cases not selected for family history cancer.

Five new p53 mutations were identified in six pedigrees from hereditary cancer clinics. No deleterious mutations were identified in cancer probands from 37 HBC/HBOC families. A targeted mutation screen of the 1,235 breast cancer cases identified a c.844C>T [p. Arg282Trp] mutation carrier. This mutation was also found among the six mutation-positive cancer families provided by the local hereditary cancer clinics.

The targeted screen also uncovered a new p53 gene mutation - c.685T>C [p Cys229Arg] - in two breast cancer cases. p53 gene mutation carriers were found in all of the 656 women with breast cancer diagnosed at less than 50 years of age.

The p53 tumor suppressor protein has emerged as a universal sensor of genotoxic stress that regulates the transcription of numerous genes required for appropriate cellular response to DNA damage. Therefore, the transcriptional induction of p53 protein target genes can be considered a global and early indicator of genotoxic stress. By performing expression microarrays and RNA-Seq analysis on wild-type and mutant p53 genes in human lymphocytes respectively, derived from controls and Li-Fraumeni patients and exposed to different classes of genotoxic agents, we first determined a common p53-dependent transcriptional signature of DNA damage.

The specificity of the p53 genotoxicity assay can easily be demonstrated by performing the same experiment in control lymphocytes with heterozygous p53 mutations, which compromise responses to DNA damage. This assay allowed us to show that most of the drugs commonly used in cancer treatment, with the exception of the microtubule poisons, are highly genotoxic.

The p53 genotoxicity assay should facilitate the measurement of the genotoxic effects of chemical and physical agents and the identification of drugs that are not genotoxic and do not expose patients to the risk of secondary malignancies, especially those with a constitutional defect in response to DNA damage, such as patients with Li-Fraumeni syndrome.

Another strong limitation to routine analysis of p53 mutations resides in the fact that many tumors contain an excess of the wild-type p53 gene as compared with the mutant gene, resulting from the presence of intact alleles in the tumor as well as in non-cancer cells (stromata, inflammatory cells and blood vessels).

Chapter Four

Isoform p53 Protein and Cancer

The function of the tumor suppressor of the p53 protein is well-highlighted in human cancer, where it is mostly present either in mutated or amyloid/aggregated form. Under normal conditions, p53 protein is inactivated by MDM2 protein (murine double minute 2), a negative regulator that causes proteasomal degradation of p53 protein.

Phosphorylation of p53 protein caused by diverse cellular stresses can reduce its binding affinity to MDM2 protein, which results in its activation. The p53 protein subsequently forms a homo-tetramer that binds to specific p53 response elements in the genomic DNA, where it acts as a transcriptional regulator of its downstream genes, regulating the cell cycle, apoptosis, DNA repair, and several other vital functions to maintain the genome's integrity.

Control of p53 protein activity is achieved by post-translational modifications, such as phosphorylation, acetylation, and ubiquitination, which influence p53 protein's binding to the DNA and also allow interaction with other proteins thereby affecting p53 protein transcriptional functions. On the other hand, isoform p53 protein exhibits a loss of its obligatory transcriptional activities, thereby gaining new functions.

Previous studies have suggested that in tumor cells, a non-functional p53 protein is often present in an aberrant, misfolded, and inactive conformation. Amyloid formation begins with the unfolding of p53 protein structural folds, followed by the formation of structures containing β-strands. Recent studies on the aggregation of p53 protein have shown its internalization into cells, its seeding capacity, and cell-to-cell transmission in a prion-like fashion. Interestingly, not only wild-type p53 but also mutant p53 proteins can also form amyloids, thereby leading to tumorigenesis.

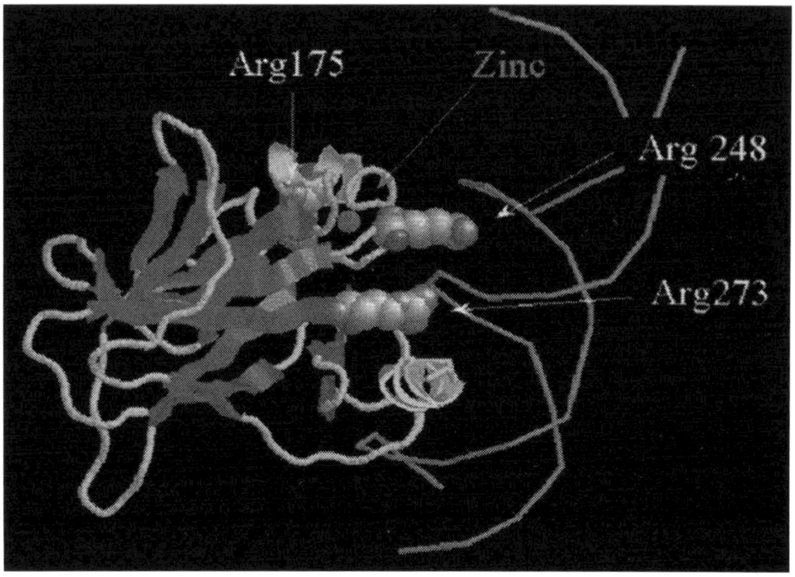

Figure 1. *Three-dimensional shapes of the x-rayed p53 protein structure in isomorphic form. Five p53 gene mutations were identified in hereditary cancer clinics. A targeted mutation screen of cancer cases identified an Arginin-282-Tryptophan mutation carrier.*

Cell lines investigated showed not only differences in the amount of total p53 protein but also displayed distinct differences in the phosphorylated forms of the protein. All tumor-derived cell lines showed p53 expression; as expected, the non-cancerous fibroblast cell line (lane #4) showed no p53 expression.

These results clearly show that p53 activity is affected in different ways in different cancers. For example, the ovarian cancer cell line (lane #3) showed increased expression of p53, activated through phosphorylation of serine 6, while the colon cancer cell line (lane #5) showed up-regulation of p53, phosphorylated at serine 15. The use of the Multi-Blot Kit greatly facilitated this analysis as all forms of p53 shown here are of virtually the same molecular weight [**Scheme 4**].

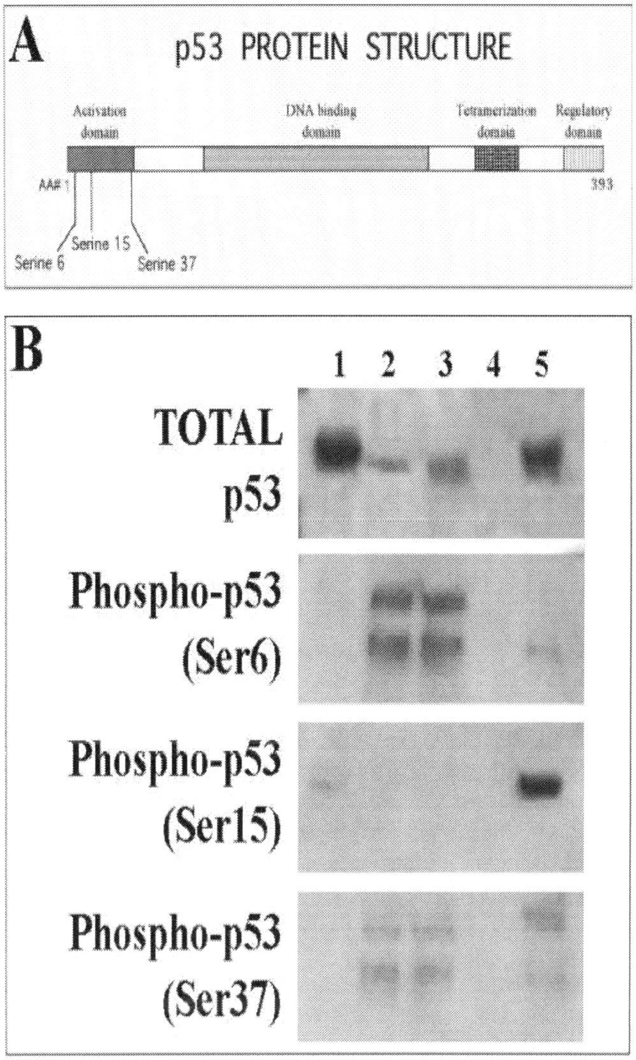

***Scheme 4**. A schematic of the known protein domains in p53. Phosphorylation of residues within the activation domain at the N-terminus of the p53 protein results in the activation of the protein. Three serine residues are associated with this activation.*

The crystal structure of p53 protein DNA binding domains (as found in the bioactive homo-tetramer) has seven domains:

1. An acidic N-terminus transcription-activation domain (TAD), also known as activation domain 1 (AD1), activates transcription factors: residues 1-42. The N-terminus contains two complementary transcriptional activation domains - a major one at residues 1–42 and a minor one at residues 55–75 - specifically involved in the regulation of several pro-apoptotic genes.

2. Activation domain 2 (AD2) is important for apoptotic activity: residues 43-63.

3. The proline-rich domain is important for the apoptotic activity of p53 protein by nuclear exportation via MAPK: residues 64-92.

4. Central DNA-binding core domain (DBD). Contains one zinc atom and several arginine amino acids: residues 102-292. This region is responsible for the binding of the p53 co-repressor LMO3 [22].

5. Nuclear localization signaling domain: residues 316-325.

6. Homo-oligo-dimerization domain (OD): residues 307-355. Tetramerization is essential for the activity of p53 in vivo.

7. C-terminal involved in downregulation of DNA binding of the central domain: residues 356-393. The International Cancer Genome Consortium has established that the p53 gene is the most frequently mutated (>50%) in human cancer, indicating that the p53 gene plays a crucial role in preventing cancer formation. The p53 gene encodes proteins that bind to DNA and regulate gene expression to prevent mutations in the genome.

The only way to study these proteins in the same sample using conventional Western blotting would be to strip and re-probe the blot multiple times. This is cumbersome, but more importantly, protein bound to the membrane is removed with each cycle of stripping. The Multi-Blot Kit allows confident comparisons of same-sized proteins within a sample. Characteristics of apoptosis are chromatin condensation, membrane blebbing, phosphatidyl-serine exposure on the cell's surface, cytoplasm shrinkage, the formation of apoptotic bodies, and DNA fragmentation **[Figure 2]**.

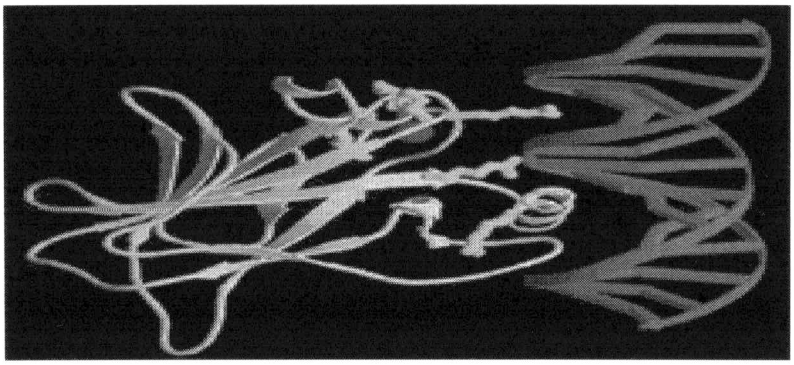

The structure of the core domain of the p53 protein (light blue) bound to DNA (dark blue) The six most frequently mutated amino acids in human cancers are shown in yellow - all are residues important for p53 binding to DNA. Red ball: zinc atom. [Reproduced from Cho, Y., et al. (1994) Science, 265, 346-355, with kind permission.]

Figure 2*. Structure of the core domain of the isoform p53 protein (yellow colour).*

The p53 protein is located in the nucleus of cells throughout the body, where it attaches (binds) directly to DNA. When the DNA in a cell becomes damaged by agents such as toxic chemicals, radiation, or ultraviolet (UV) rays from sunlight, this protein plays a critical role in determining whether the DNA will be repaired or the damaged cell will self-destruct (undergo apoptosis). If the DNA can be repaired, p53 protein activates other genes to fix the damage.

If the DNA cannot be repaired, this protein prevents the cell from dividing and signals it to undergo apoptosis. By stopping cells with mutated or damaged DNA from dividing, p53 helps prevent the development of tumors.

Given the key role that anti-apoptotic proteins play in CLL, they have become an attractive target for the creation of novel therapeutic agents, such as the anti-sense methodology and small molecules mimicking the action of the BH3 domain and microRNAs (mainly miRNA-15a and miRNA-16-1). For several decades, the front-line therapy for CLL has been represented by treatments with alkylating agents and purine analogues.

Recently, significant clinical outcomes have been achieved using chemo-immunotherapy, such as rituximab, a chimeric monoclonal antibody against the CD20 receptor, administrated in combination with fludarabine or fludarabine plus cyclophosphamide. Nevertheless, relapse remains problematic, particularly in older patients, thus the identification of innovative and specific agents against CLL remains of high interest.

An interesting development is the use of a cottonseed extract derivative gossypol, which acts as a BH3-mimetic, interfering with the functions of Mcl-1, Bcl-2 and Bcl-xL proteins and displacing pro-death partners to induce apoptosis. However, despite encouraging data and abundant literature describing the molecular mechanisms triggered by phytochemicals to inhibit cell growth and induce apoptosis in cancer cells, only a few of them entered clinical trials [9].

CHAPTER FIVE

MUTANT P53 GENES AND HEMATOLOGIC MALIGNANT DISEASES

p53 gene mutations are the most common genetic abnormalities of cancer. They have been extensively studied in various malignant diseases of mature B cells, including CLL. In the latter, most attention has been paid to the importance of p53 protein expression in CLL and a combination of low survival rates and non-responses to classical conventional chemotherapy and radiotherapy has been observed in several CLL cases.

Mutation of the p53 gene affects the cell cycle and cells lose control of cell proliferation, leading to the propagation of damaged DNA into their progenies, which become transformed into cancerous cells. Isoform p53 proteins might either act as a dominant negative over wild-type p53 action or gain novel tumorigenic properties counteracting the protective function of wild-type p53 proteins.

Nuclear p53 phosphoprotein regulates normal cell proliferation and the accumulation of non-functional mutated p53 genes in tumor cell nuclei is associated with the development and/or progression of several neoplastic diseases, including breast cancer.

Activation of p53 protein induces senescence and cell-cycle arrest under excessive oncogenic stress and this is a crucial mechanism for regulating the inhibition of tumorigenesis.

Cell proliferation is induced by inactivating the p53/p21 Cdk-interacting protein 1 (Cip1) complex via Ras-dependent and -independent stimulation of the Raf/MEK/ERK cascade. These findings point to the importance of Ras oncogenic activity and p53 protein inactivation in human cancers.

The p53/p21/p27 and p53/Bcl-2/ Bax pathways affect many biological processes including cell proliferation, the G2/M phase and apoptosis. Specific chromosomal abnormalities and genetic mutations can serve as diagnostic and prognostic indicators for disease progression and survival. The effectiveness of new therapies should be tested for the presence of these molecular lesions in CLL patients.

Recent researches have shown that the p53 gene is a tumor suppressor gene and its activity inhibits tumor formation. In the cell, p53 nuclear protein binds to DNA by stimulating another gene, CDKN1A, which produces a protein called p21 that interacts with a cellular division-stimulating (CDK2) protein. In this context, it has been shown that the nuclear p53 protein protects the cell from a malignant process and the cytoplasmic protein p53 through its isoforms can gain new functions to promote carcinogenesis **[Scheme 5]**.

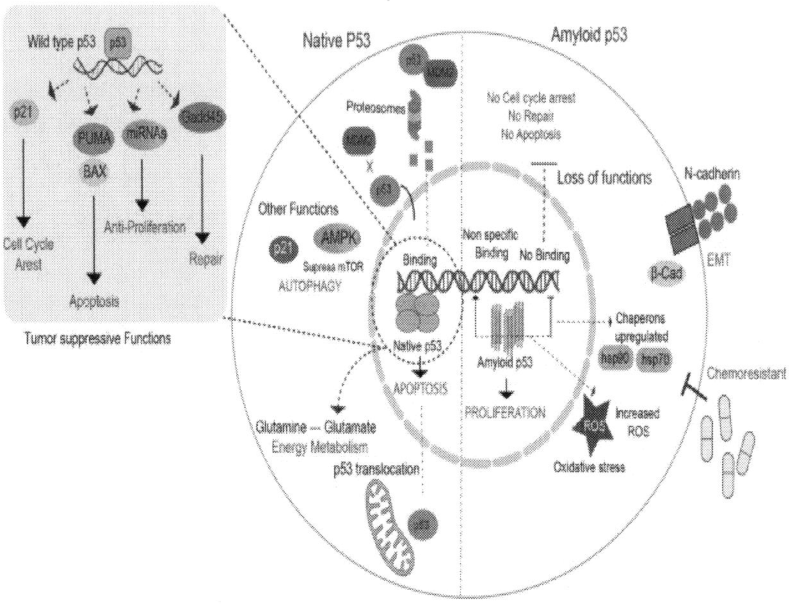

Scheme 5. *The modified activity of the p53 protein in isoforms affects DNA damage and extends from the light to the severe phenotypes of cancer.*

Acetylation, deacetylation and phosphorylation of the p53 protein are reversible and can repair DNA damage in cancer cells. The spectrum of cancer phenotypes due to mutations in the p53 gene is also supported by the fact that different isoforms of the p53 protein have different cellular mechanisms in cancer.

These findings suggest that phosphorylation of the p53 protein at the serine-15 amino acid is an important focal point in p53 protein activation. The amino acid replacement, serine alanine, caused the partial failure of p53 protein to inhibit cell cycle progression. Also, the p53 protein has been identified as an important regulator of glucose transport, and the transcriptional repression of both GLUT1 and GLUT4 receptors has been demonstrated. By contrast, the p53 mutant does not affect GLUT1 and GLUT4 receptor activity in the malignant cell.

Other studies have suggested that a high adenosine triphosphate (ATP) concentration in CLL B-cell malignancies has affected the p53 gene and induced cell apoptosis. It is recommended that before initiating treatment in CLL, ATP levels are correlated with the presence or absence of p53 gene mutations and should be measured to adjust the doses of cytostatic treatments that are resistant, causing the relapse of the disease. To avoid very high doses of cytostatic, which may cause adverse effects, recurrences from CLL are also recommended for the administration of ATP inhibitor analogue drugs, e.g. isopentenyl ester of ATP (Appl), which is a triphosphoric acid ester and inhibits cell division cycle triggering signals.

It has been shown that in CLL, p53 gene mutations are commonly detected in two alleles of chromosome 17p, occurring in more than 75% of cases. Patients with both abnormalities have a significantly increased resistance at treatment than those with just a single deleted allele 17p.

Some research has shown that another mechanism that can trigger p53 gene dysfunction is the overexpression of the MDM2 gene, which involves suppressing a large number of p53-dependent genes and mRNAs, including microRNA-34a. Since this microRNA is involved in the induction of apoptosis depression and cell cycle disruption, a more aggressive course of the disease can be correlated with the excessive expression of microRNA 34 [10, 11].

Chromosomes were also associated with the increased intensity of CD20, FMC7 and CD79b receptors on the B-cell surface. In addition, there is an increase in CD38, ZAP-70 and IGHV expression in 17p derelict cases.

Furthermore, mutations in tumor suppressor genes, such as the p53 gene and ATM, have been associated with resistance to conventional chemotherapeutic agents.

Modifications of micro-RNA expression and aberrant methylation patterns in genes that are specifically deregulated in CLL, including the BCL-2, TCL1 and ZAP-70 genes, have also been encountered and related to distinct clinical parameters.

In addition, an increased expression of CD38 receptors, ZAP-70 protein kinases and non-GM IGHV chains has been reported in 17p deletions, resulting in poor prognoses for this group of ZAP70 patients in CLL.

In contrast to wild-type p53 protein with the anti-tumor protective activity, isoform p53 proteins have oncogenic action in culture cells, and promote metastasis and genomic instability in mice models. Mutations of the p53 gene are often alterations in the central DNA-binding domain and several hotspots such as R175, G245, R248, R249, R273 and R282, have been so far identified. In both cases, the structural stability of p53 protein is altered and isoform p53 protein might acquire a gain/loss-of-function phenotype **[Scheme 6]**.

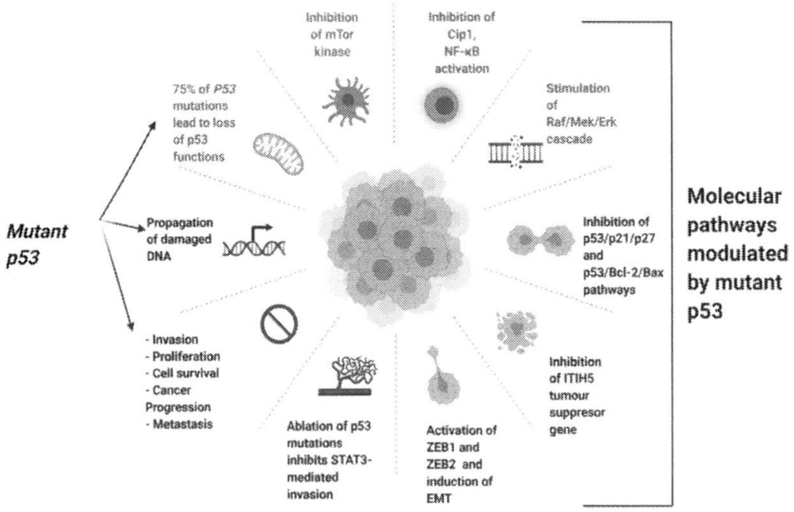

Scheme 6. *Protein p53 mutations are subdivided into two main categories structural and DNA-contact mutations, which affect either the folding of the p53 protein or the transcriptional activity of p53 and regulation of target genes.*

The interaction of isoform p53 proteins with non-canonical/different response elements might induce oncogenic responses; indeed, isoform p53 proteins might act as tumor-initiating transcriptional factors. Isoform p53 proteins may also control gene expression through different mechanisms in which they interact directly with DNA sequences that bind to nuclear matrix regions, giving another way to change/regulate gene expression.

Post-transcriptional modifications of mutant p53 proteins have a rapid turnover due to ubiquitination mediated by the E3 ubiquitin ligase MDM2 protein and subsequent proteasomal degradation. p53 protein stability in response to genotoxic stress is regulated by a variety of post-translational modifications (PTMs), including phosphorylation, acetylation, methylation, glycosylation, ubiquitination, and methylation, which occur in different regions of the p53 protein, and play a crucial role in regulating p53's stability and localization, thereby affecting p53's ability to modulate cell proliferation and cell death.

Phosphorylation of p53 and its role in cancer progression and apoptosis regulation in different serine (S) and threonine (T) phosphorylation sites have been identified, in particular in the C- and N-terminal domains of p53 proteins. Notably, serine 15 phosphorylation is pivotal in the activation of p53, as it enhances p53 stability by promoting its dissociation from the MDM2 ubiquitin ligase.

ATM (ataxia telangiectasia mutated) and ataxia-telangiectasia-mutated-and-Rad3-related (ATR) protein kinases phosphorylate p53 at serine 15 and activate and improve the stability of p53 by enhancing the interaction between p53 and the histone/lysine acetyltransferase (HATS). ATM- and ATR-mediated phosphorylation of downstream substrates, including casein kinase (CK1), checkpoint kinase 1 (Chk2), and p53, limit cancer cell viability by modulating many critical biochemical pathways leading to cell cycle arrest, DNA repair, senescence, and death.

Phosphorylation of p53 on other serine residues activates/ modulates p53 functions, and is associated with cell exposure to different toxic or damaging stimuli, such as UV and IR. DNA damage after UV and IR exposure mediates the stabilization of human p53 through phosphorylation on Ser-20. Apoptosis is triggered instead by p53 phosphorylation on Ser-46, which is mediated by many kinases, including homeodomain-interacting protein kinase 2 (HIPK2), p38, and dual-specificity tyrosine-phosphorylation regulated kinase 2 (DYRK2).

Genotoxic stress causes the transactivation of p53 by modulating p53 phosphorylation on its amino terminus, which regulates the interactions between p53 and MDM2 orp300/CBP. The interactions of p53 with the apoptosis-stimulating proteins of p5 (ASPP) proteins, as well as p300, control its apoptotic activity.

Acetylation of p53 is another post-translational reversible enzymatic process by which p53 action is fine-tuned in response to different cellular toxic signals, including genotoxic stress and DNA damage.

Many studies have looked at the role that acetylation plays in regulating p53 as a tumor suppressor. p53 proteins are acetylated at defined amino acid positions.

In particular, lysine (K) residues within the C-terminal regulatory domain (K370, K372, K373, K381, K382, and K386) of p53 are key acetylation sites and play a crucial role in regulating p53 functions **[Scheme 7]**.

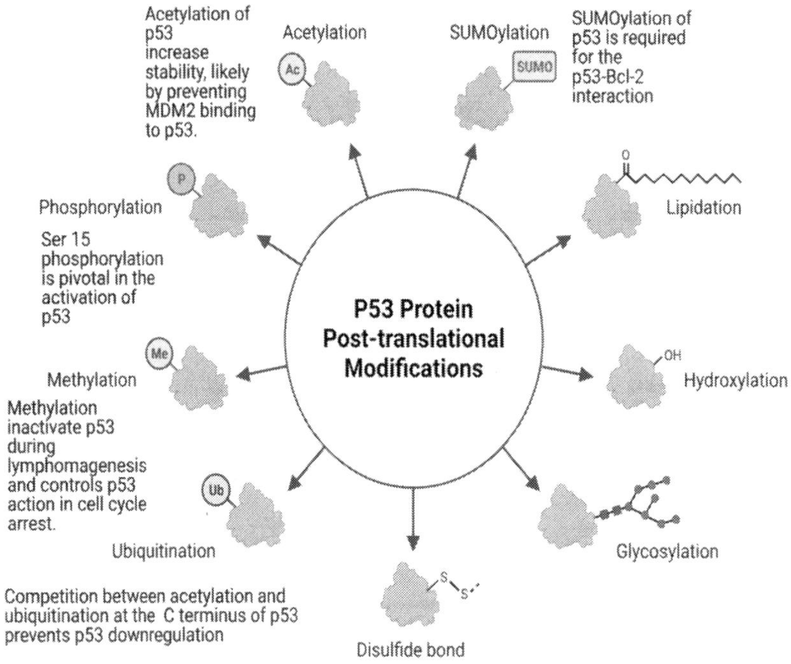

Scheme 7. *Phosphorylation, acetylation, methylation, glycosylation, and ubiquitination occur in different regions of the p53 protein and play a crucial role in regulating p53 stability and localization in response to genotoxic stress.*

Also, p53 protein ubiquitination mostly occurs at its C-terminus, where acetylation also occurs during cell stress, and by competing with ubiquitination it prevents p53 downregulation.

These data support a connection between accumulating missense p53 mutations and NF-κB activation in human cancers linked with colitis. The combined binding of NF-κB, the R273H p53 mutant, and other mutant versions of p53 to these enhancers regulates RNA polymerase II recruitment to these elements in colorectal carcinomas, boosting mRNA synthesis and the activation of tumor-promoting genes.

The mutant p53 gene, in combination with NF-κB, can therefore alter the inflammatory tumor microenvironment (TME), inducing in both epithelial and nonepithelial cells the expression of the cancer-promoting gene. Using 9-aminoacridine derivatives in renal cell carcinomas and small molecule curtains in several cancer cell lines and mouse tumor xenografts to restore wild-type p53 function is a rational approach that has previously been demonstrated. In the mutant p53 background, approaches aimed at restoring wild-type p53 expression might potentially improve existing NF-κB-dependent therapies.

Chapter Six

Autophagy, a Process Mediated by the p53 Protein, as a Prospective Strategy for Cancer Therapy

Autophagy is a cellular process involved in the degradation of subcellular components that can be a source of proteins involved in tumorigenesis. Despite its role in maintaining genomic stability, numerous studies indicate a tumor-supportive function that allows tumor cells to respond to stress stimuli such as nutrient deficiency and hypoxia, thus extending the life of the cell. Furthermore, tumors with autophagic deficits are more susceptible to several chemotherapeutic agents.

In this case, autophagy promotes the survival of cancer cells and protects them from the action of drugs that induce apoptosis. Adjusting autophagy is quite complicated. This involves a series of signaling cascades, including the p53 protein. Recent reports have shown that p53 protein plays a dual role in regulating autophagy, depending on its sub-cellularity.

Many cancer cells contain inactivating mutations of the p53 gene, which explains why cancer cells continue to live. Under these conditions, p53 protein has been shown to mediate genes that induce autophagy and stimulate autophagy by inhibiting the AMPK-activated AMPK (AMPK) protein kinase mTOR complex.

The p53 protein is becoming an attractive therapeutic target for anticancer drug discovery. Three classes of p53-targeting compounds have been identified and characterized.

The first class is comprised of compounds that activate or restore wild-type p53 function and may be used in human cancers. These include Nutlin-1, a cis-imidazole analogue. The second class is made up of PKI-587, a dual competitive ATP inhibitor that targets PI3K (PI3K-α and

PI3K-γ) and mTOR, and that is an orally bioavailable inhibitor that rescues the p53 protein isomorph carrying a mutation of the p53 gene.

The third class, the heterocyclic compound EYE-354, is a bi-ATP-competitive kinase inhibitor that selectively blocks mTORC1/2 activity. High glucose levels increase demand for the molecular building blocks needed to support the rapid proliferation of cancer cells through the Warburg effect.

Reciprocally, glucose maintains mutant p53 gene stability and promotes cancer cell growth by generating a positive regulatory loop. The mutant p53 gene, supported by glucose to stimulate cell proliferation, defines a unique point of vulnerability in cancer cells. This glucose appetite identifies a potential therapeutic target that is under extensive investigation (i.e., ketogenic diets and the use of the drug metformin).

p53 protein has additional roles, which may overlap with its tumor-suppressive capacity, in processes including DNA damage response, metabolism, ageing, stem cell differentiation and fertility.

Consistent with this role, p53 activity is compromised in a large proportion of all cancers, either by mutation of the p53 gene (p53 protein encoding) or by modification of the p53 gene's modulator status. Phosphorylation of the N-terminal end of the p53 protein by a specific protein kinase disrupts the MDM2-binding protein and activates the p53 protein.

Stimulation of hypoxic inductor factor, HIF-1 and HIF1- target genes by mutant p53 protein results in autophagy inhibition. Thus, autophagy defects impair the survival of apoptosis-defective tumor cells upon nutrient and oxygen limitation, leading to cell death by necrosis, which in turn is associated with inflammatory cell recruitment, cytokine production and nuclear factor Kb (NFκB) activation, which has been linked to accelerated tumor growth **[Scheme 8]**.

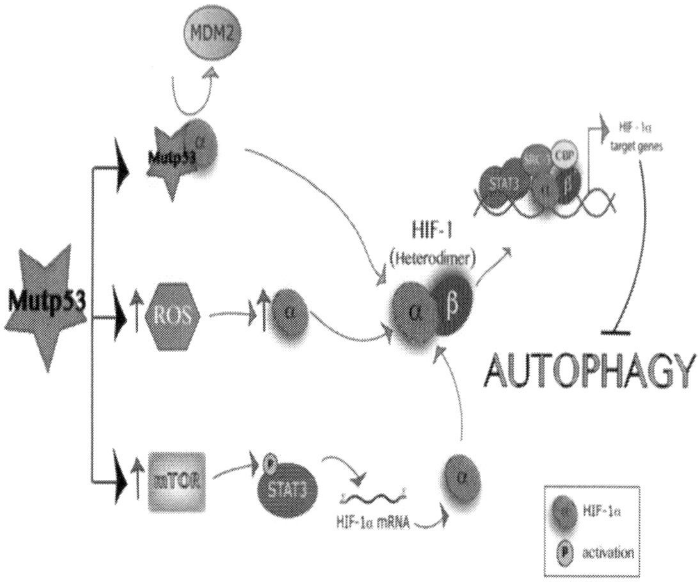

Scheme 8. *Schematic representation of the mechanisms at the basis of the stimulation of HIF-1 and HIF-1a target genes by mutant p53 protein resulting in autophagy inhibition.*

Under these conditions, native or normal p53 protein has been shown to transactivate autophagy-inducing genes and stimulate autophagy by inhibiting mTOR and activating AMP-activated protein kinase (AMPK) that stimulate the ULC Complex and Beclin 1, factors that trigger autophagy. The circumstances and the molecular pathways involved in the decision to use p53 for autophagy activation versus inhibition in cancer cells have not yet been determined.

Many cancer cells contain inactivating mutations of the p53 protein, which explains why those cancer cells go on living. Plausibly, p53 protein loss, and thus autophagy induction, or negative regulation of autophagy inhibition, may be one of the compensatory mechanisms that tumors cells use to counterbalance the survival of autophagy suppression by an activated PI3K/AKT/mTOR axis **[Scheme 9]**.

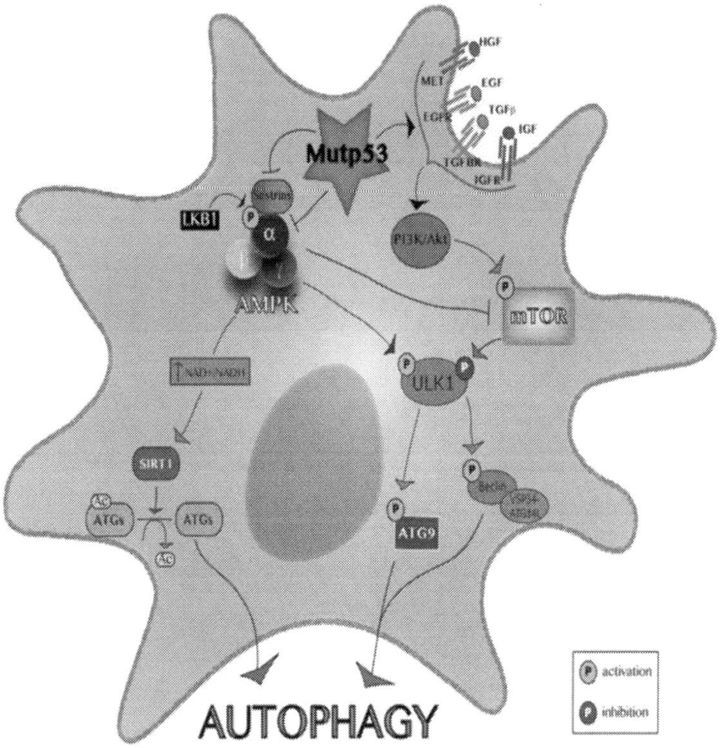

Figure 9. *Schematic representation of the molecular mechanisms through which mutant p53 protein inhibits AMPK enzymes and stimulates mTOR complex enzyme pathways, blocking autophagy.*

The mechanism by which autophagy defects lead to accelerated tumorigenesis is not readily apparent, especially given the well-documented pro-survival function of autophagy, which prolongs both normal and tumor cell survival under metabolic stress. The protein kinases that are known to target this transcriptional activation domain of p53 protein can be roughly divided into two groups. The first group of protein kinases belongs to the MAPK family (JNK1-3, p38), which is known to respond to several types of stress, such as membrane damage, oxidative stress, osmotic shock, and heat shock. A second group of protein kinases (ATM, CHK1, CHK2, DNA-PK and P53RK) is implicated in genome growth **[Scheme 10]**.

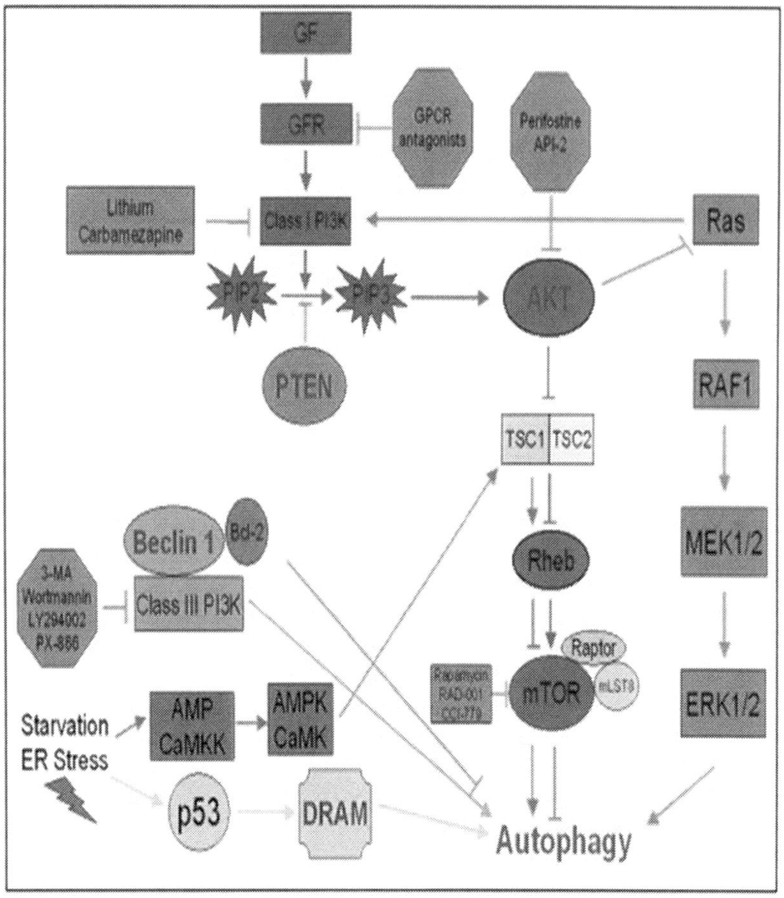

***Scheme 10. Phosphatase and** tensin homolog (PTEN) is a tumor suppressor protein that regulates the enzyme AKT.*

Two pathways are responsible for autophagy activation in response to starvation and stress: a) mediated by AMPK and Ca-MKK; and b) involving p53 and Damage-Regulated Autophagy Modulator (DRAM) activation.

Constitutive activation of the PI3K/AKT/mTOR axis is a prototypic survival mechanism commonly encountered in human cancer. Diverse cellular events, such as the loss of the tumor suppressors like the phosphatase and tensin homolog (PTEN), deletions on chromosome 10, amplification or mutation of class I PI3K and overexpression of AKT, constitutive activation of tyrosine kinase growth factor receptors and exposure to carcinogens, can all result in abnormal activation of this pathway.

Several mutations or the loss of PTEN function affect the activity of lipid phosphatase, which leads to the development of a variety of cancers. Of the three residues in the PTEN component, R-335 was found to be the most important in interactions with the cellular membrane, in common with several other germline mutations, and was associated with inherited cancer.

Autophagy defects are associated with susceptibility to metabolic stress, DNA damage, accumulation, genomic instability, and accelerated tumorigenicity. These observations lead to the predictions that autophagy stimulation may preserve cellular fitness and genome integrity and thus prevent cancer, and that tumors with chronic autophagy deficiency may be particularly sensitive to certain anticancer agents, such as DNA-damaging and anti-angiogenic drugs.

Ultimately, pharmacologic manipulation of autophagy for cancer prevention and treatment will depend on our ability to successfully recognize the functional status of autophagy in tumors and on the availability of specific autophagy modulators.

Consistent with this role, p53 protein activity is compromised in a large proportion of all cancers, either by mutation of the p53 gene (p53 protein encoding) or by modification of the p53 protein's modulator status. The contextual role of autophagy in cancer, which could be changed by p53's status, is expected to be developed into a new anticancer therapeutic approach. Nuclear p53 protein facilitates autophagy by transactivating its target genes, whereas cytoplasmic p53 protein mainly inhibits autophagy through extranuclear, transcription-independent mechanisms.

1.1 Molecular Mechanism of Autophagy

Autophagy is a degradation pathway characterized by the isolation of targeted cytoplasmic material in a typical double-membrane vesicle, known as an autophagic vacuole or autophagosome. The subsequent fusion of the autophagosome with the lysosome ensures the correct destruction of the organelles, misfolded proteins and microorganisms carried inside the vesicle.

Despite its emerging role in human pathology, autophagy is a physiological process involved in the turnover of basal organelles and the removal of protein aggregates. In response to cellular stress conditions, such as growth factors and nutrient deprivation, intracellular components degraded by autophagy are recycled to generate ATP and sustain essential cell functions.

Autophagy is considered a pro-survival mechanism, allowing cells to respond to injury by degrading unnecessary and dysfunctional self-components; however, this ability may become a double-edged sword. Three types of autophagy can be distinguished: macro-autophagy, micro-autophagy, and chaperone-mediated autophagy. In this review, we will focus on macro-autophagy (hereafter referred to as autophagy), which is the most characterized type of autophagy [11].

Considering the crucial role of autophagy in the maintenance of cellular homeostasis, it is not surprising that several signaling-related molecules are involved in the perfect functioning of this process. Genetic screens in yeasts allowed the discovery of at least 37 autophagy-related genes (ATGs). Many of these genes, encoding proteins involved in autophagy and its regulation, are evolutionarily conserved in humans.

The mammalian target of rapamycin (mTOR) complex 1 (mTORC1) regulates the activation of the autophagy machinery, acting as a sensor of energy levels and integrating upstream signals derived from other pathways, including the phosphoinositide 3-kinase (PI3K)-Akt **[Scheme 11]**.

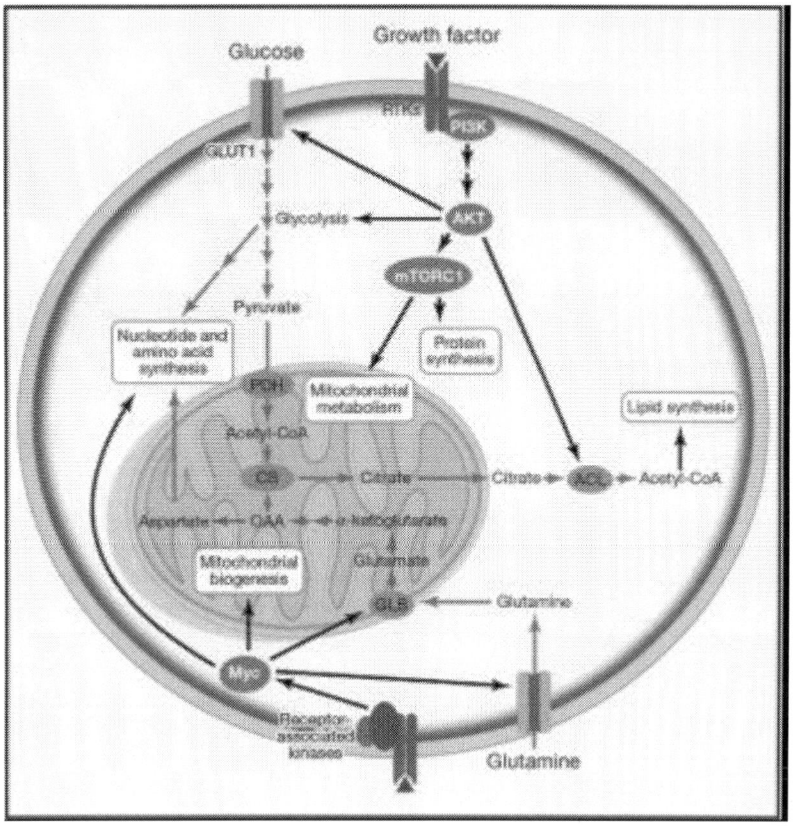

Scheme 11. *Mitochondrial signals by the PI3K-Akt- mTORC1 pathway.*

In the presence of amino acids and growth factors, mTORC1 represses autophagy by the inhibition of Vps34 and ULK1 complexes. On the contrary, in low-nutrients state, defined as starvation, the dissociation of mTORC1 from the induction complex triggers autophagy [12, 13, 14].

The autophagosome derives from a double-membrane pre-autophagosome structure called the phagophore, which seems to originate from different sources, including the plasma membrane endoplasmic reticulum (ER) and the Golgi complex in mammalian cells. Phagophore nucleation requires the activity of a class III phosphatidylinositol 3-kinase (PI3K-III) complex containing Beclin-1 [15].

The initial signal to form autophagosomes is from the class III phosphatidyl-inositol (PI) 3 kinase complex that consists of the sequence genes Beclin1/Atg6, p150hVSp35, and class III PI3K (Vps34).

The autophagy-promoting function of Beclin-1 is influenced by the antiapoptotic protein Bcl-2: when Beclin-1 is bound to Bcl-2, autophagy is inhibited, but dissociation from Bcl-2 allows Beclin-1 to interact with the PI3K-III complex and to activate autophagy.

On the other hand, the Beclin-1-regulated autophagy protein 1 (AMBRA1) is a positive regulator of Beclin-1-dependent autophagy, thanks to its capacity to create a link between cytoskeletal motor proteins and the class III PI3K complex. Two ubiquitin-like conjugation systems, Atg12–Atg5–Atg16L and microtubule-associated protein 1 light-chain 3 (LC3)-phosphatidylethanolamine (PE), mediate the second step of autophagy, which concerns the expansion and closure of the autophagosome. In the first system, the enzymes E1-like Atg7 and E2-like Atg19 promote the covalent association of Atg12 to Atg5 (**Scheme 12)**.

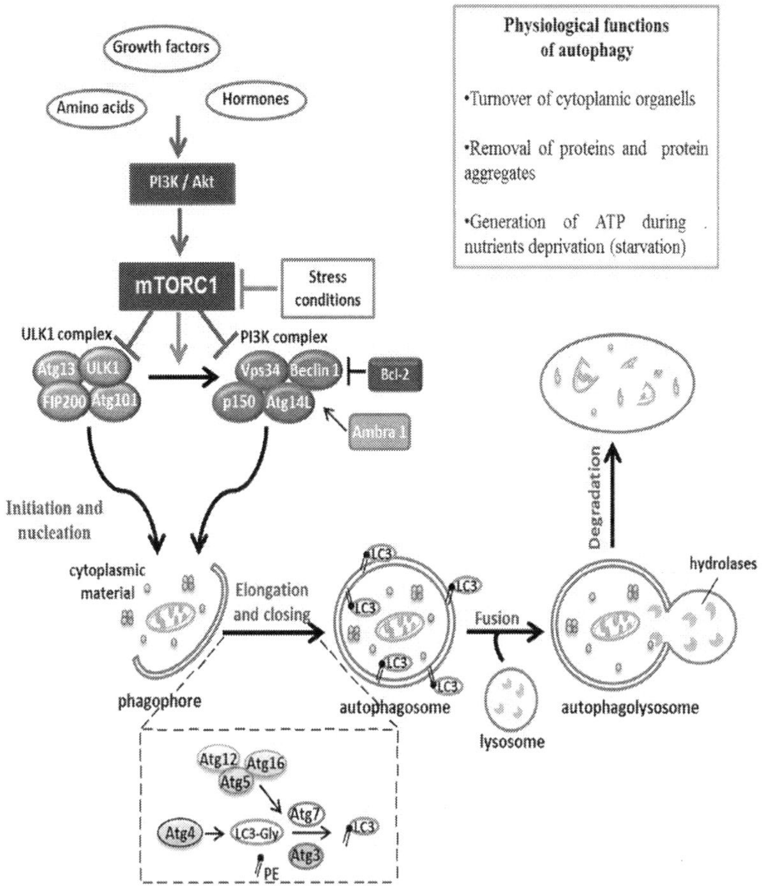

Scheme 12. *Schematic representation of signaling pathways involved in autophagy multistep regulation. Autophagy represents a fundamental physiological process, considering its role in ATP production during the condition of nutrient deprivation and in controlling organelles and protein turnover. mTOR complex 1 (mTORC1), acting as a sensor of energy levels and integrating upstream signals [phosphoinositide 3-kinase (PI3K) and Akt], is one of the most important autophagy regulators in cells. In the presence of growth factors and hormones, mTORC1 inhibits autophagy activation. The fusion of the autophagosome with the lysosome exposes the cargo to the action of lysosomal hydrolases that degrade vesicular content for recycling.*

1.2 p53 protein-mediated autophagic: A prospective strategy for cancer therapy

The relationship between autophagy and neoplasia is complicated. It may be intrinsically associated with the functional status of the p53 protein, but this has not been elucidated. p53 protein has been studied intensively as a major tumor suppressor that detects oncogenic events in cancer cells and eliminates them through senescence (a permanent non-proliferative state) or apoptosis. Consistent with this role, p53 protein activity is compromised in a high proportion of all cancer types, either through mutation of the p53 gene (encoding p53) or changes in the statuses of p53 protein modulators.

Phosphorylation of the N-terminal end of the p53 protein by specific protein kinases disrupts MDM2-binding **[Scheme 13]**.

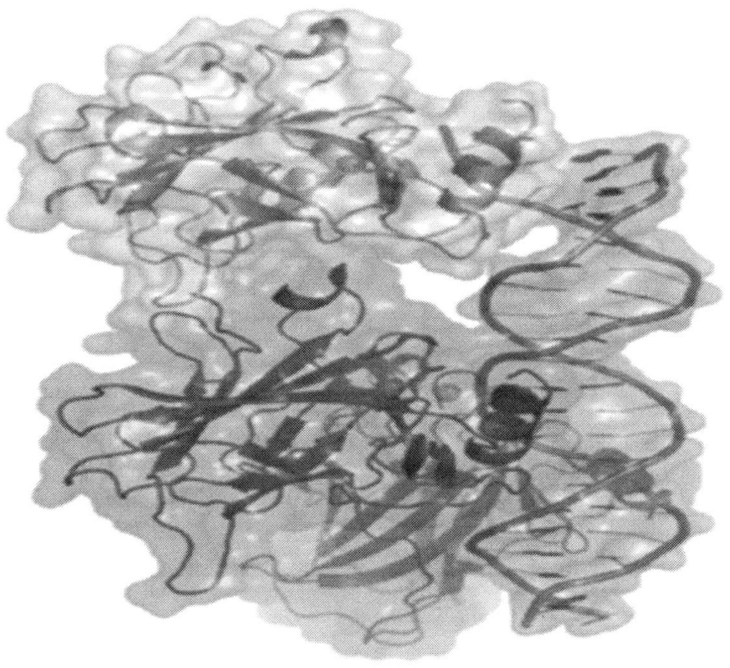

Scheme 13. *Crystal structure of four p53 DNA binding domains (as found in the bioactive homo-tetramer).*

The protein kinases that are known to target this transcriptional activation domain of p53 can be roughly divided into two groups. The first group of protein kinases belongs to the MAPK family (JNK1-3, ERK1-2, and p38 MAPK), which is known to respond to several types of stress, such as membrane damage, oxidative stress, osmotic shock, heat shock, etc. A second group of protein kinases (ATR, ATM, CHK1, CHK2, DNA-PK, CAK, and P53RK) is implicated in the genome integrity checkpoint, a molecular cascade that detects and responds to several forms of DNA damage caused by genotoxic stress.

Oncogenes also stimulate p53 activation, mediated by the protein p14ARF. In carcinogenesis, a series of signaling cascades including the p53 protein are involved.

Recent reports have indicated that p53 protein plays a dual role in regulating autophagy depending on its subcellular localization. Nuclear p53 protein facilitates autophagy by transactivating its target genes, whereas cytoplasmic p53 protein mainly inhibits it. The contextual role of autophagy in cancer is expected to be developed into a new anticancer therapeutic approach.

Additionally, p53 protein activity is regulated through critical context-specific or fine-tuning events, mediated primarily through post-translational mechanisms, particularly multi-site phosphorylation and acetylation. The present review has broadly examined these events, highlighting their regulatory contributions, their ability to integrate signals from cellular events towards providing the most appropriate response to stress conditions, and their importance for tumor suppression [16].

The p53 protein has been studied for many years because it is a tumor suppressor. Many cancer cells contain inactivating mutations of the p53 protein, which explains why those cancer cells go on living. The tumor suppressor p53 protein is a critical checkpoint protein in mammalian cells, and is activated under genotoxic stress conditions including DNA damage, hypoxia and oncogene activation, and responds by initiating tumor suppression mechanisms, such as cell cycle arrest, senescence and apoptosis. Under these conditions, the p53 protein has been shown to transactivate autophagy-inducing genes and stimulate autophagy by inhibiting mTOR in an AMP-activated protein kinase (AMPK)- and TSC1/TSC2-dependent manner.

At the same time, however, genetic or pharmacologic inactivation of cytoplasmic p53 also triggers autophagy, indicating that the non-nuclear p53 pool is a potent autophagy repressor.

Thus, autophagy is activated as a stress-mitigating mechanism by both stress-mediated p53 induction and stress-exacerbating p53 protein loss. The circumstances and the molecular pathways involved in the decision to use p53 for autophagy activation versus inhibition in cancer cells have not yet been determined.

Plausibly, p53 protein loss, and thus autophagy induction, or negative regulation of autophagy inhibition, may be one of the compensatory mechanisms that tumor cells use to counterbalance the survival-undermining effects of autophagy suppression by an activated PI3K/AKT/mTOR axis [17].

Regulators of apoptosis, such as Bcl-2/Bcl-xL and the BH3-only proteins, interact with Beclin 1 and can modulate autophagy. The anti-apoptotic protein Bcl-2 binds to Beclin 1 under non-stress conditions and inhibits autophagy in the ER, whereas the BH3-only protein BAD, BNIP3 and BH3 mimetics such as ABT737 competitively inhibit the interaction between Beclin 1 and Bcl-2/Bcl-xL and stimulate autophagy.

Thus, positive regulators of apoptosis also induce autophagy, which is not very surprising given that both pathways are activated under similar stress conditions. The functional relationship between autophagy and apoptosis is complicated by the fact that there is an apparent hierarchy in the activation and execution of these two stress-responsive pathways.

On the other hand, when autophagy is blocked, apoptosis is accelerated or apoptosis-defective cells undergo metabolic catastrophe and die by necrosis. Thus, cell fate in response to metabolic stress is determined by the functional status and the interaction between the stress-mitigating pathways of apoptosis and autophagy.

Regarding cancer prevention, the function of autophagy as a protector of cellular homeostasis and genome integrity may be particularly important.

Autophagy inhibition as a means to sensitize cancer cells to treatment has been validated in several studies: inhibition of autophagy by chloroquine, a lysosome-tropic agent that raises intra-lysosomal pH and interferes with autophagosome degradation within lysosomes, was shown to enhance the anticancer activity of the alkylating agent cyclophosphamide in a Myc-gene-induced lymphoma model and to induce p53-dependent cell death and tumor suppression in different Myc-gene-induced and ATM-deficient lymphoma models.

1.3. Regulation of autophagy in cancer

Increasing evidence has revealed that autophagic dysfunction is associated with human diseases, such as cancer. Paradoxically, although autophagy is well recognized as a cell survival process that promotes tumor development, it can also participate in a caspase-independent form of programmed cell death. Induction of autophagic cell death by some anticancer agents highlights the potential of this process as a cancer treatment modality.

Dysfunctional autophagy contributes to many diseases. Large-scale genomic analysis of human cancers indicates that the loss or mutation of core autophagy genes is uncommon, whereas oncogenic events that activate autophagy and lysosomal biogenesis have been identified. Autophagic flux, however, is difficult to measure in human tumor samples, making a functional assessment of autophagy problematic in a clinical setting. Autophagy impacts cellular metabolism, the proteome, and organelle numbers and quality, which alter cell functions in diverse ways [18].

Moreover, autophagy influences the interaction between the tumor and the host by promoting stress adaptation and suppressing the activation of innate and adaptive immune responses. Additionally, autophagy can promote a cross-talk between the tumor and the stroma, which can support tumor growth, particularly in a nutrient-limited microenvironment. Thus, the role of autophagy in cancer is determined by nutrient availability, microenvironment stress, and the presence of an immune system.

Defective autophagy is implicated in tumorigenesis, as the essential autophagy regulator Beclin 1 is mono-allelically deleted in the human breast, ovarian and prostate cancers, and Beclin 1+/− mice are tumor-prone. Cell-autonomous mechanisms, involving the protection of genome integrity and stability, and a non-cell-autonomous mechanism, involving suppression of necrosis and inflammation, have been discovered so far. Autophagy inhibition concurrent with chemotherapy or radiotherapy has emerged as a novel approach in cancer treatment, as autophagy-competent tumor cells depend on autophagy for survival under drug- and radiation-induced stress.

A lower level of Beclin 1 protein expression, as compared to Beclin 1 levels in normal adjacent breast tissue, was confirmed in a small series of human breast tumors but any correlation between allelic Beclin 1 loss, and thus defective autophagy, and clinical outcome in breast cancer remains to be investigated. The mechanism by which autophagy defects lead to accelerated tumorigenesis is not readily apparent, especially given the well-documented pro-survival function of autophagy, which prolongs both normal and tumor cell survival under metabolic stress [19].

Beclin 1 was independently rediscovered in a yeast two-hybrid screen of an adult mouse brain library as a Bcl-2 interacting protein. Beclin 1 maps to a centromeric region of BRCA1 on chromosome 17q21 that is commonly deleted in 75%, 50% and 40% of ovarian, breast, and prostate cancers, respectively.

In addition to Beclin 1, this commonly deleted region contains at least 11 more genes, 6 of which are of known function and not considered cancer related, and 5 of which are completely novel genes. Thus, autophagy defects impair the survival of apoptosis-defective tumor cells upon nutrient and oxygen limitation, leading to cell death by necrosis, which in turn is associated with inflammatory cell recruitment, cytokine production and nuclear factor κB (NFκB) activation, which has been linked to accelerated tumor growth.

1.4. PI3K/AKT/mTOR pathway in autophagy

Constitutive activation of the PI3K/AKT/mTOR axis is a prototypical survival mechanism commonly encountered in human cancer. Diverse cellular events, such as the loss of the tumor suppressor protein PTEN, the tuberous sclerosis complexes TSC1 and TSC2, amplification or mutation of class I PI3K, overexpression of AKT, constitutive activation of tyrosine kinase growth factor receptors, and exposure to carcinogens, can all result in abnormal activation of this pathway and, ultimately, in autophagy suppression.

The phosphatase and tensin homolog (PTEN) is a tumor suppressor protein that regulates the enzyme phosphoinositol-3 kinase (PI-3K). Several mutations or loss of PTEN function affect the activity of lipid phosphatase, which leads to the development of a variety of cancers. Of the three residues in the PTEN component, R-335 was found to be most important to interact with the cellular membrane, in common with several other germline mutations, and was associated with inherited cancer [20] **[Scheme 14]**.

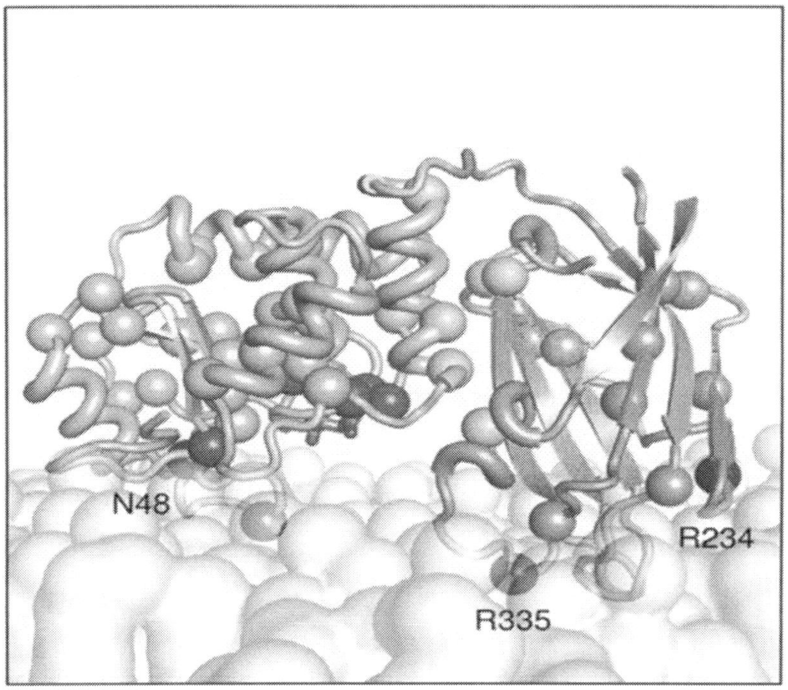

***Scheme 14**. Phosphatase and tensin homolog (PTEN) in membrane location interacting in clinically important mutations.*

On the contrary, G-protein coupled receptor (GPCR) antagonists to growth factor receptors (GFR), class I PI3K inhibitors such as lithium and carbamazepine, AKT inhibitors such as perforin and AKT/PKB-signaling inhibitor-2 (API-2), and mTOR inhibitors such as rapamycin, RAD-001 and CCI-779, result in autophagy induction and provide multiple means for autophagy modulation in cancer treatments **[Scheme 15]**.

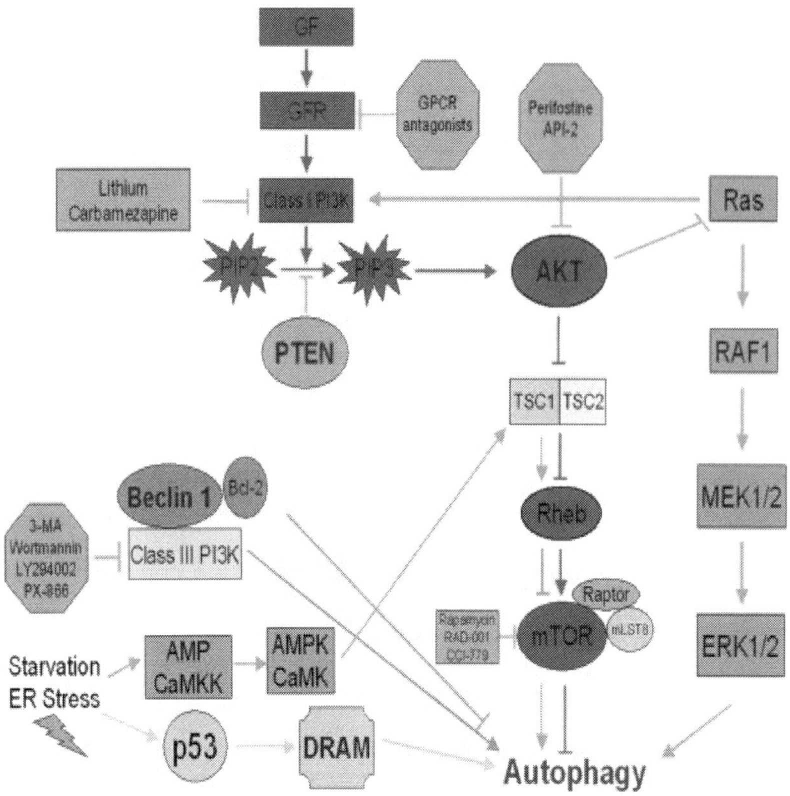

Scheme 15 *Autophagy regulation with growth factor signaling which activates the PI3K/AKT/mTOR axis, resulting in autophagy inhibition (move pathway).*

At least two pathways are responsible for autophagy activation in response to starvation and ER stress: one mediated by AMPK and Ca-MKK, and the other involving p53 and damage-regulated autophagy modulator (DRAM) activation. In this complex regulatory network, Ras exhibits a dual function as both an autophagy inhibitor (via class I PI3K activation) and an autophagy activator (via the RAF1/MEK1/2/ERK1/2 pathway).

Tumors with high metabolic demands, such as those with constitutively active PI3K mutations, PTEN loss or AKT activation, would be expected to be dependent on autophagy for energy homeostasis and survival. Thus, the suppression of autophagy by the PI3K signaling cascade presents a disadvantage that these rapidly proliferating tumor cells may have to overcome to remain viable and leads to the prediction that compensatory mechanisms, such as deregulated apoptosis and/or metabolism, may be concurrently activated to counteract the negative implications of defective autophagy on tumor cell survival [21].

1.5. Autophagy in immunological tolerance

Many studies have demonstrated autophagy's contribution to the presentation of cytosolic antigens in association with MHC class II molecules, playing an important role not only in the acquired immune response but also in the maintenance of self-tolerance. Mechanisms of central (in the primary lymphoid organs) and peripheral tolerance (in peripheral tissues) physiologically prevent immune responses to self-antigens.

During T cells' development in the thymus, the recognition of peptide–MHC molecules on the surface of thymic epithelial cells (TECs) ensures that only thymocytes restricted to MHC molecules, and specific for non-self (foreign) antigens, will survive and continue their maturation. Emerging evidence indicates that autophagy contributes to the maintenance of the central tolerance mechanism.

It was recently revealed that there had been an alteration in the selection of the T cell receptors (TCRs) restricted to MHC class II in mice with transplanted Atg5−/− thymuses. Autophagy defects, in association with a consequent loss of self-tolerance, could be the reason for the multiple signs of autoimmunity reported in these animals. These two opposing results probably depend on the different approaches used to inhibit autophagy in the thymus, thus further investigations are necessary.

The involvement of autophagy in the presentation of self-antigens to immature T cells in the thymus was first analyzed by Kasai and colleagues, who showed a colocalization of LC3-II with the lysosomal compartment in which MHC–peptide complexes are formed.

More recently, Aichinger et al. (2020) demonstrated that autophagy is essential for endogenous antigen-loading onto MHC class II of TECs for negative selection.

1.6. Autophagy in lymphocyte homeostasis

Peripheral immune cells play an important role in the perpetuation of autoimmunity by sustaining systemic inflammation status and by participating in the extension of joint destruction mechanisms. Many studies have demonstrated that autophagy allows T and B lymphocytes to survive in conditions of nutrient deprivation or during stress stimuli.

Mice lacking Atg5 do not survive and have a reduction of peripheral T cells, showing how autophagy is essential for their survival. Since cytoplasmic calcium levels are essential for the activation of TCR-signaling pathways, autophagy-dependent calcium flux regulation could influence T lymphocyte activation. It has been demonstrated that CD4+ and CD8+ Atg5−/− cells are not able to properly proliferate following TCR stimulation.

Moreover, the inhibition of autophagy causes defects in T-cell activation. Deletion of Atg7 results in a decrease in the IL-2 mRNA level and ATP generation, suggesting that autophagy is required to ensure an appropriate energy level for T cell activation. Similar data were also obtained on B lymphocytes, demonstrating that autophagy is essential for the maturation process and for the subsequent maintenance of the B lymphocyte repertoire in the periphery.

1.7. Reconciling effects of autophagy on malignant transformation and tumor progression

Autophagy has been attributed to tumor-suppressive as well as tumor-promoting functions, raising doubts about the actual therapeutic value of autophagy inhibitors for cancer therapy. Autophagic responses almost invariably exert beneficial effects, but these are detrimental for the host when they occur in cells that have already initiated malignant transformation.

Moreover, the true impact of autophagy on oncogenesis and tumor progression may exhibit a significant degree of context dependency. It is critical to characterize specific neoplasms for their actual dependency on autophagy before implementing the use of autophagy inhibitors in the clinical routine [22].

There is no doubt that the role of autophagy in cancer is complex and that much work is still needed to determine the molecular details of autophagy regulation in tumor cells, define how the functional status of autophagy in tumors impacts cancer progression and response to treatment, and elucidate how to best modulate autophagy for cancer prevention and therapeutic benefit.

Autophagy defects are associated with susceptibility to metabolic stress, DNA damage accumulation, genomic instability, and accelerated tumorigenicity, and these observations lead to the predictions that autophagy stimulation may preserve cellular fitness and genome integrity, and thus prevent cancer, and that those tumors with chronic autophagy deficiency may be particularly sensitive to certain anticancer agents, such as DNA-damaging and anti-angiogenic drugs.

Tumors with intact autophagy likely utilize and potentially rely on this pathway for survival under metabolic stress conditions, such as during rapid tumor growth, metastasis and chemo- or radiotherapy. In this case, inhibition of autophagy concurrent with treatment may augment the antitumor activity, and thus the efficacy, of radiation and/or anticancer drugs. Ultimately, pharmacologic manipulation of autophagy for cancer prevention and treatment will depend on our ability to successfully recognize the functional status of autophagy in tumors and on the availability of specific autophagy modulators [23].

CHAPTER SEVEN

CHRONIC LYMPHOCYTIC LEUKEMIA (CLL), A SHORT REVIEW

Chronic lymphocyte leukemia (CLL) occurs in the middle-aged and elderly, affecting men and women at a rate of about 2:1. Many patients are asymptomatic when the disease is diagnosed. Patients with a minimal expression of the disease, i.e. only with numerous lymphocytes, are considered to be in the incipient phase of the disease, while those who demonstrate compromised functioning of the bone marrow, with anemia or thrombocytopenia, were in the advanced stages of the disease.

The cause of CLL is unknown. There is no link to radiation. It is unclear if certain chemicals can cause CLL. CLL usually affects older adults, especially those over the age of 60. People under 45 rarely develop CLL. CLL is more common in whites than in other ethnic groups. It is more common in men than in women. Some people with CLL have family members with the disease. Tests to diagnose CLL include: complete blood count (CBC) with blood cell differential, bone marrow biopsy, CT scan of the chest, abdomen, and pelvis, immunoglobulin testing, and flow cytometry test of the white blood cells.

The CLL diagnosis can be initially established by optical microscopy and lymphocyte morphology combined with immunophenotype: monoclonal antibodies in CD5+ panel receptors, CD19+, CD20+ and CD23+, CD28 with lymphocytes by expressing heavy chains of immunoglobulins IgM or IgD type IgM Kappa or Lambda chains. Hypo-gama-globunlinemia is the first sign of the disease in most patients with LLC. Different biological and genetic markers also have prognostic values in the CLL. Chromosomal abnormalities can be detected by FISH technology in up to 80% of LLC patients.

Most clonal abnormalities involve the long arm q of chromosome 13, and band 13q14, with good prognostics, trisomy 12, deletions 11q22-q23 and

7q21-q23, and deletion 17p13. These are associated with more advanced stages of the disease or shortened survival times.

Chronic lymphocytic leukemia (CLL) is a lymphoproliferative disorder characterized by small, mature monoclonal CD5+/CD19+ B cells which continue to accumulate in the peripheral blood (PB), bone marrow (BM) and lymphoid organs.

There is remarkable clinical variability in patients with CLL2. Following diagnosis, some patients have asymptomatic disease that may not progress for many years. Others are diagnosed with advanced-stage disease or have an early-stage disease that progresses rapidly and requires treatment. Clinical symptoms include peripheral blood lymphocytosis, lymphadenopathy and hepatosplenomegaly. When the disease progresses, many patients develop hypogammaglobulinemia and neutropenia, resulting in increased susceptibility to bacterial infections.

Furthermore, up to 25% of the patients develop autoimmune cytopenia, mostly autoimmune hemolytic anemia and/or autoimmune thrombocytopenia.

This abnormality is found in 10% of patients with CLL and has a pure prognosis. Deletion of the long arm of chromosome 13(del 13q) is the most common genetic abnormality in CLL with roughly 50% of patients exhibiting this effect.

These patients have the best prognosis and most will live many years without the need for therapy. Agents damaging DNA may increase the expression of p53 and its trans-activation activity, suggesting that p53 acts to protect cells.

The clinical variability is associated with the following molecular markers: cytogenetic abnormalities [3]; immunoglobulin VH gene mutational status [4; 5]; ZAP-70 expression [6; 7]; and CD38 expression [8]. How these factors influence the biology of CLL is still the subject of intensive research. Immunoglobulin gene mutational status and ZAP-70 expression are associated with the capacity of CLL cells to respond to signals delivered through the B-cell receptor (BCR) [9]. Loss of 17p [10] or 11q [11] may cause dysfunction of p53 DNA damage pathways [12].

Classically, CLL cells were thought to derive from naive B lymphocytes and to behave as inert cells that passively accumulate [13]. However, mRNA studies showed that at the level of gene expression, both IgVH mutated and unmutated CLL cells were most similar to memory B cells [14, 15].

Moreover, stable isotope labelling studies with deuterated water (D2O) have shown that CLL cells divide at a considerable rate, suggesting a more dynamic disease than previously appreciated [16]. Survival and proliferation of the CLL cells are influenced by interactions with non-leukemic cells in the microenvironment of lymph nodes (LN), bone marrow and spleen [1, 17, 18]. Still, deregulated apoptosis is considered to be a key factor and a major barrier to the effective treatment of CLL.

Understanding apoptosis pathways and the impact of the microenvironment on these pathways is therefore clinically relevant, because it may open new avenues to the effective treatment of CLL [24].

This is the most common leukemia in the Western world, primarily affecting the elderly, and is characterized by the accumulation of mature, circulating, IgM-low CD5+ B cells. Several genetic aberrations with prognostic value and impact on treatment decisions in CLL have been described. These include deletions of the chromosomal regions 17p13 (containing the p53 suppressor gene), 11q23 (containing the DNA damage checkpoint protein ATM), or 13q14 (miR-15a, miR-16-1), and trisomy of the chromosome.

Furthermore, > 80% of cases harboring del(17p) also carry p53 gene mutations in the remaining allele. Such patients with p53 gene defects are classified as 'high-risk' and often respond poorly to therapy. Moreover, a significant proportion of CLL patients carry a p53 gene mutation in the absence of a 17p deletion.

Given that ibrutinib shows off-target inhibition of JAK3, ITK and EGFR, it can be used to target oncogenic pathways other than BTK in tumor cells and as a T-cell modulator in combination immunotherapy.

Treatment with ibrutinib increased cellular persistence and decreased expression of co-inhibitory surface molecules on chimeric antigen receptor (CAR) T cells in models of CLL and MCL. Whether ibrutinib acts on ITK in CAR T cells or on BTK in the malignant cells or other kinases remains undetermined.

Paradoxically, inhibiting ITK in T-cells may be efficacious in cancer, as it may enhance the Th1-skewing of CD4+ T-cells and thereby improve the memory formation and functionality of CD8+ T-cells, potentially leading to improved anti-tumor immunity. These potentially beneficial off-target effects of ibrutinib may be lost in the highly specific BTK inhibitors that are currently being evaluated.

The targeting of BTK, which has a central role in several signaling pathways in B-cells, particularly in the BCR, has shown impressive efficacy as a therapeutic option for various B-cell malignancies in clinical trials. Much progress has been made in recent years in defining the complex mechanisms of action of BTK inhibition.

These involve intrinsic signaling pathways in leukemic cells that are central to cellular survival, proliferation and, most importantly, retention in supportive microenvironments. Moreover, BTK inhibition shows promise as a therapy that influences crucial immune cells in the tumor microenvironment.

Because data from BTK-deficient or inhibitor-treated myeloid cells in the context of cancer are scarce, it is not clear whether BTK inhibition by, for example, ibrutinib is based on its specificity for BTK in particular myeloid cells.

As ibrutinib treatment in CLL cases diminished the immunosuppressive properties of malignant cells through BTK-dependent and BTK-independent mechanisms (probably via ITK inhibition, it will be interesting to observe whether the same level of anti-tumor efficacy is maintained by specific BTK inhibition alone. It is conceivable that for particular malignancies it may be advantageous to use BTK inhibitors that show additional specificity for related kinases.

Mitogen-activated protein kinase (MAPK) cascades have been shown to play a key role in the transduction of extracellular signals to cellular responses. In mammalian cells, three MAPK families have been characterized: classical MAPK (also known as ERK), C-Jun N-terminal kinase/stress-activated protein kinase (JNK/SAPK) and p38 kinase. MAP kinases lie within protein kinase cascades.

The stimulation of tyrosine kinase receptors (RTKs) provokes the activation of MAPKs in a multi-step process. For example, the essential links from epidermal growth factor receptors to MAP kinases include adaptor protein Grb2, a guanine nucleotide exchange protein, such as SOS, a small GTP binding protein, p21ras, a cascade of protein kinase defined sequentially as MAPKKK (represented by c-Raf-1), and MAPKKs such as MEK1 and MEK2. MEKs ultimately phosphorylate p44 MAPK and p42 MAPK, also known as ERK1 and ERK2 respectively, thereby increasing their enzymatic activity.

Activated Ras or MEK proteins were shown to induce the expression of reporter genes driven by the cyclinD1 promoter [13]. Terada et al. demonstrated that the cyclinD1 promoter contains two potential sites targeted by the activity of the Ras/Raf function. The cyclinD1 promoter's activity increased significantly when a constitutively activated form of MKK1(S222E) was expressed and inhibited by the MKK1 inhibitor PD9805914. The c-Jun response element might be important for the expression of the cyclinD1 protein and the responsive element might be a mediator for the normal growth factor.

Given the dependence of the cyclinD1/Cdk4 function on Rb, the Ras function in mid-to-late G1 is Rb-dependent. The JNK signal transduction pathway is implicated in multiple physiological processes. Three genes encode JNK (α, β and γ) with 12 possible isoforms derived from alternative splicing products [22].

Several MAPKKKs have been reported to activate the JNK signaling pathway. These include members of the MEKK group. JNK activation is associated with transformation in many oncogenes and growth factor-mediated pathways. The transactivation of c-Jun might play an important role in this process. JNKs can transduce signals for differentiation in the hematopoietic system, and possibly involve in embryonic development. JNK pathway has been implicated in both apoptosis and survival signaling.

It has been reported that UV-induced apoptosis in fibroblasts requires JNK for the release of cytochrome C from the mitochondria [25].

Recently, numerous reports have indicated the MAP kinase pathways are involved in many pathological conditions, including cancer and other diseases. It seems that MAP kinase signaling pathways represent a potential target for therapeutic intervention. Therefore, a better understanding of the relationship between the MAP kinase signal transduction system and the regulation of cell proliferation is essential for the rational design of a novel pharmacotherapeutic approach.

CHAPTER EIGHT

CLINICAL AND PATHOPHYSIOLOGICAL FEATURES OF CLL

The objective of this study was to present the latest research in the field of molecular medicine, in terms of CLL, that has emerged with regard to the p53 gene with deletions or translocations in the human lymphoma genome, and the prognoses and treatments of this disease. I also wanted to discover whether the following were correlations between characteristics of LLC disease and response to treatment with tyrosin-kinase inhibitor treatment until the indication of allogenic stem transplantation.

In recent years, fluorescence in situ hybridization (FISH) has been proven the best technique in the investigation of malignant lymphocytes. The method is a gene analysis technique (recombinant DNA technology) and consists of coupling fluorescent-labelled nucleic acid probes with a specific chromosomal region (in situ hybridization with digoxigenin or fluorescent classical dye, molecular genetics and genomics).

The method is used to identify chromosomal abnormalities and their numerical and structural sites. The principle of this method consists of attaching to the target sequence a single-stranded DNA probe (about 40 kb) fluorescently labelled based on the complementary with a target sequence of a chromosome.

FISH-engineering can detect the chromosome deletions 7q, 13q, 11q, 13 q, 14q and 17p from peripheral blood leukocytes or bone marrow. Hybridization of the probe with the cellular DNA is visualized in the fluorescence microscope equipped with excitation and emission filters, which enables reading the target as a specific signal.

1.1. Previous researches result in a genetic panel of Chronic Lymphocytic Leukemia

In a previous international study based on 109 cases of CLL, the majority of patients, 67%, had a single anomaly, and the remaining 33% had two or three genetic abnormalities.

In another study, 90 cases of CLL were analyzed for the presence of the lymphocytic membrane receptor of B lymphocytes, CD38, and the 81 cases were placed in a group prognosis. Nineteen (23%) of the 81 were CD38 + with a poor prognosis. In a similar percentage of cases CD38 + was present with cromozomiale deletions 17q and 11q (33%) and translocation (4q-11q (36%)), while cases resulting from FISH normal were in the 33% percentage.

CLL cases with trisomy 12 or isolated 13q- had the lowest percentage of CD38 + cases: 15% and 8%, respectively. A ZAP-70 receptor was tested in 36 cases; 10 were positive.

1.2 Genetic markers discovered by FISH technology

Deletions of the chromosomal region 13q14 were frequently associated with cases of CLL and lymphocytosis with monoclonal B-cells may occasionally precede LLC or aggressive lymphomas and it is suggested that this region contains a tumor suppressor gene, similar to the p53 gene, which combines their work. All this leads us to conclude that the appearance of LLC deficiency occurs through a common p53 gene and the chromosomal region 13q14 gene.

CLL and Hodgkin's lymphoma (HL) are particularly dependent on their microenvironments and have the associated signaling pathways and deletion of miR15/16 loci that are common in CLL. Micro-ARN, miR15 and miR16 are located at chromosome 13q14, a region deleted in more than half of B-cell chronic lymphocytic leukemia (B-CLL) cases. The micro-RNA deletions, miR-15a and -16-1, both accelerate the proliferation of human B-cells by modulating the expression of genes that control cell cycle progression. These results define the role of the 13q14 chromosome with deletions of the p53 gene in the pathogenesis of CLL.

Another study used FISH to detect cytogenetic features in 275 cases of B-CLL in 48 hospitals in China. The frequencies of the abnormalities were as follows: deletion 13q (56.4%), trisomy 12 (34.5%), deletion of the p53 gene (33.5%) and 11q22 deletions (30.5%). It was evident that patients with p53 gene deletion had lower levels of hemoglobin ($P = 0.001$) and PLT ($P = 0.003$) compared to patients without the gene deletion p53.

In general, somatic mutations in the gene p53, due to the influence of physical factors or chemical or biological infections over the years, tended to predict clinical outcomes in cancer patients due to the central role of the p53 tumor suppressor gene.

The role of adhesion molecules such as integrins and E-cadherin (CDH1) in the signal transduction of intracellular cancer cells has been elucidated above.

The presence of contact cell-cell and cell-matrix contacts play an important role in the modulation patterns of morphological and gene expression of cancer cells.

The extracellular matrix (ECM) of a tumor microenvironment is composed of various glycoproteins and fibrous proteins that allow cancer cells to interact with each other and with normal cells. The characteristic surface roughness of the edge of the spheroid SNU-387 was much more prevalent as compared to the SNU-475 cell line. Adjusting FISH with immune-fluorescence staining of the interphase cell lines showed the p53 protein, the product of the p53 gene, with varying intensity and color CDH1 **[Photo 1]**.

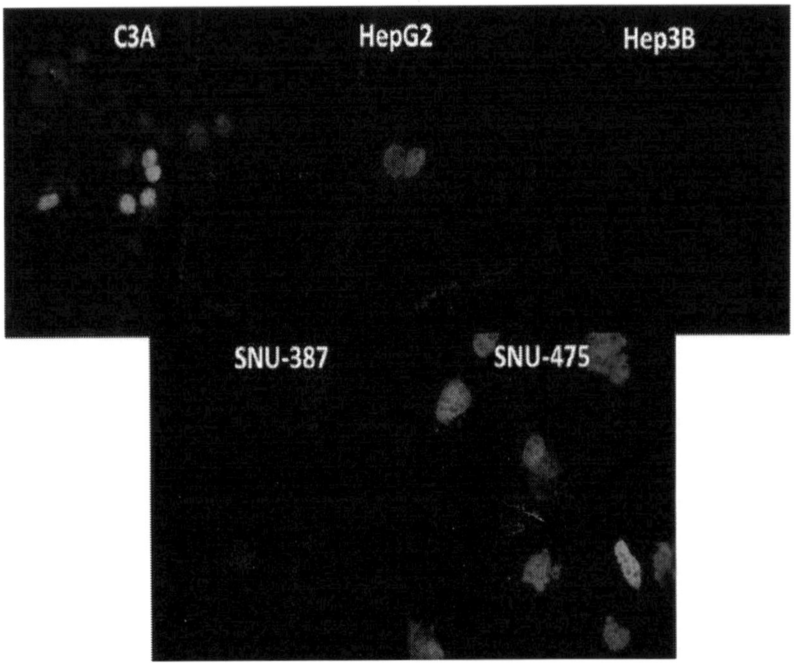

***Photo 1**. For wild-type p53 cell lines, HepG2 and C3A presented intensities with intermediate levels of staining in the nucleus of each cell. [K. Sabapathy, Contribution mutant p53 proteins in oncogenesis. Sci. Republication beyond 2015 (5). Email: gs.moc.sccn@bskrmc.com. Picture published with author's agreement 1/10/2016 for Aurelian Udristioiu, at his request.]*

In the other conditions, a recent paper systematically explored the optimal conditions required for the growth of tumor spheroids in hepatocarcinoma. Also, researchers evaluated gene expression patterns of the p53 gene and E-cadherin [CDH1] in selected cell lines. CDH1 is located on the surface of cancer cells and plays an increasingly important role in the formation of cell-cell contacts and is responsible for the maintenance of cell polarity. In addition, CDH1 plays an important role in the physiology of the extracellular matrix (EMT) **[Photo 2]** [26].

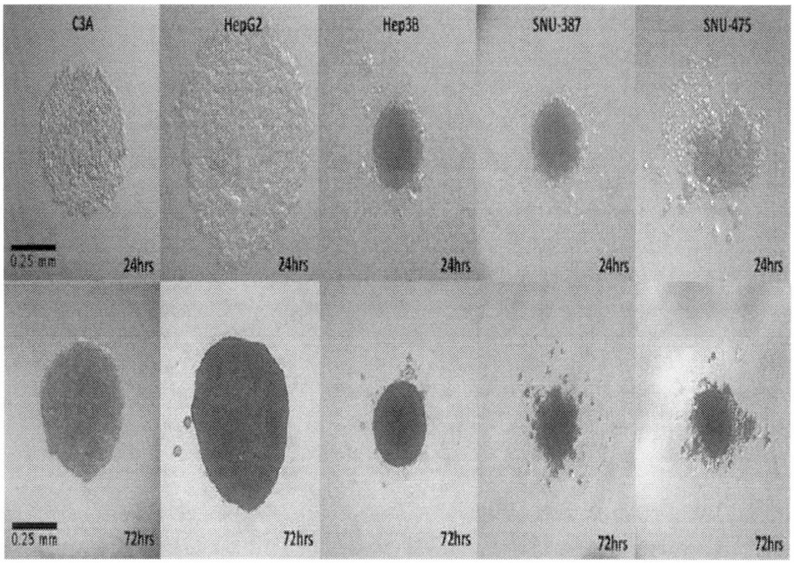

Photo 2*. Influence of p53 and CDH1 genes in hepatocellular cancer spheroid formation of culture. [Picture published with author`s agreement 1/10/2016 for Aurelian Udristioiu, at his request.]*

In particular, isoform p53 protein, as a result of p53 gene mutation, is known to exhibit a dominant-negative effect, which can lead to an abnormal accumulation in the nucleus. However, the dominant-negative effect typically depends on the specific position of the isoform p53 protein.

In normal cells, the suppressor p53 gene, coding proteins that bind to DNA and regulate the expression of genes, prevents genome mutations. A mutation of the p53 gene will inevitably lead to a process of carcinogenesis, in which the cell divides endlessly.

Identification of p53 gene mutations in the region of the 17 chromosomes of hematologic neoplasm is important because these mutations have an impact on the clinical course of patients and require an attitude adjustment therapeutic adequate.

However, the primary pathogenic event that leads to the growth, proliferation and survival of B-cells in CLL is difficult to determine. Molecules involved in B-cell receptor (BCR) signaling pathways constitute a rise in cytoplasmic survival factors, acting in concert to confer resistance to apoptosis.

1.3. Biological activities of the isoform p53 protein

Cancer genome sequencing studies confirm that the p53 gene is the most commonly mutated tumor suppressor gene in human cancers. The majority of studies indicate that the presence of a mutated p53 gene is associated with a bad prognosis in various cancer types. Five new p53 mutations were identified in six pedigrees from hereditary cancer clinics (17p13.1).

Mutations of the p53 gene are known primarily to inactivate properties of the onco-suppressor p53 protein wild type, losing the function of transcription (loss-of-function [LOF]), and this can arm cancer cells with new oncogenes (gain-of-function [GOF]).

When the p53 gene exerts its function, glucose uptake is enhanced by a GOF-producing shunt glucose transporter, Glut-1, on the surface of the cancer cell membrane.

p53 mutations are not present in normal cases and this induces exposure to a genotoxic allele. Therefore, midway mutant p53 co-exists with wild-type (WT) p53, before losing the wild-type allele by loss-heterozygote (LOH).

Moreover, mutant p53 protein, called "gain-of-function" (GOF), acquires new activities that help cancer aggression. Induction of p53 action is carried out largely by decoupling the p53-MDM2 interaction, leading to increased levels of p53.

1.4. Isoform p53 protein and cancer

Research has shown that this restoration of function of the protein p53 can result in the regression of certain cancer cells without damaging other cells. For example, restoration of the function of endogenous p53 can induce apoptosis in lymphoma while cell growth may be reduced to normal levels. Thus, the pharmacological reactivation of p53 is presented as a viable treatment option for cancer.

Acetylation of p53 protein followed by phosphorylation is an important means of post-translational modification and is indispensable for its activation which is a reversible enzymatic process in cancer. The broad spectrum of phenotypes of the p53 gene mutations in cancer is supported by the fact that the p53 protein isoforms have different cellular mechanisms of the carcinogenesis process.

Both acetylation and deacetylation of the p53 protein are involved in the regulation of cellular responses to DNA damage and genotoxic stress in a large spectrum of cancer phenotypes due to mutations in the p53 gene, as is supported by the fact that different isoforms of p53 proteins have different cellular mechanisms in cancer. Mutations in the p53 gene can give rise to different isoform p53 proteins involving functionality disturbance in different cellular mechanisms and thereby extending the cancer phenotype from mild to severe.

At the molecular level, certain stimuli, such as genotoxic stress (DNA damage-inducing agents) and glucose deprivation, promote a series of reversible post-translational modifications (PTMs) of p53, including multisite phosphorylation of the transactivation domain (N-terminus). In addition to contributing towards the induction of p53, these events are thought to regulate p53-mediated transcription at individual promoters, possibly in a selective manner in the 'barcode' hypothesis.

Ser-15 phosphorylation also triggers a sequential series of additional phosphorylation events in p53 (including phosphorylation of Ser-9 -20, -46 and Thr-18) that contribute further to p53 induction and activation. These findings suggest that Ser-15 phosphorylation is therefore a major focal point in the activation of p53. Ser-15 phosphorylation is required to permit local acetylation of histones and relaxation of chromatin.

The replacement of Ser-15 with alanine results in partial failure of the isoform p53 protein to inhibit cell cycle progression. In this context, the nuclear p53 protein has been shown to protect cells from malignant processes, and only cytoplasmatic p53 protein, by its isoforms, in a modified cytoplasmatic medium, by the high concentration of anaerobic ATP, drives cancer.

A current study has showed that the level of p21 is strongly correlated with the activity of the mammalian target of rapamycin (mTOR) but not p53 status.

This study was published in the February 2, 2016, online edition of the Nature journal (www.cnio.es). Through the Warburg effect, glucose maintains stability and the mutant p53 gene promotes cancer cell growth and generates a positive regulatory loop. This appetite for glucose in the cancer cell points to a potential therapy for malignant diseases, which is currently under extensive investigation.

p53 protein plays an important role in the regulation of glycolysis, which is shown experimentally. Most researches seem to indicate that, in line with its role as a tumor suppressor, p53 can decrease glycolysis. This activity can be regarded as an attempt to counteract the acquisition of p53 aerobic glycolysis, usually associated with cancer cells.

Of major concern, the p53 protein has been identified as an important regulator of glucose transport, and repressing transcription has been shown to both receptors GLUT1. By contrast, the mutant p53 does not affect GLUT1 and GLUT4 receptor activity.

Mutations in the p53 gene can give rise to different isoforms of protein p53, preventing their overview in different cellular functions.

Previous studies have suggested that expression of the gene p53 works like a watch and this has a circadian rhythm. Disruption of the circadian rhythms appears to be associated with accelerating the development of cancer. Current models are also useful for modelling protein isoforms of p53 mutations and their oscillation effects could promote pharmacological drug discovery of chemotherapy in the treatment of CLL.

A mutant p53 protein will no longer effectively bind DNA, and, as a consequence, the p21 protein will not be available to act as the "stop signal" for cell division [38]. The expression of the CDK gene, which encodes protein p21, is tightly controlled by the tumor suppressor protein p53, through which this protein mediates the p53- protein-dependent cell cycle G1 as well as phase arrest in response to a variety of stress stimuli [39, 40].

1.5. The effect of Aurora-kinase A and B

Aurora kinases - enzymes A and B - play critical roles in regulating spindle assembly, chromosome segregation, and cytokinesis to ensure faithful segregation of chromosomes during the mitotic cell division cycle. Aberrant expression of Aurora kinases, on the other hand, causes defects in mitotic spindle assembly, checkpoint response activation, and chromosome segregation, leading to chromosomal instability. Aberrant expression of the Aurora kinases–p53 protein family signaling axes appears to be critical in the abrogation of the p53 protein family-mediated tumor suppressor pathways frequently deregulated during the oncogenic transformation process.

Phosphorylation of p53 serine-06 has been shown to inhibit the interaction of p53 with MDM2 and prolong the half-life of the p53 protein.

Aurora-kinase B has also been shown to interact with and phosphorylate p53 protein at multiple residues in the DNA-binding domain. Similar to the effect of Aurora-A phosphorylation on p53 activity and stability, Aurora-B phosphorylation of p53 at serine-269 and threonine-284 inhibits p53 transactivation activity, whereas phosphorylation at serine-183, threonine-211, and serine-215 accelerates the degradation of p53 through the poly-ubiquitination-mediated proteasome pathway [MDM2].

Identification of p53 gene deletions and mutations in the regions of chromosome 17 in hematologic malignancies is important because these mutations have an impact on the clinical management of patients and require an attitude adjustment therapeutic adequate in personalized medicine. Personalized treatments will be applied by combining diagnostic tools, knowledge databases and therapeutic drugs. Only 2-5% of patients with chronic CLL exhibit a T-cell phenotype. In CLL, the DNA of B lymphocyte cells is mutated so that it can't fight infection, but it grows out of control and crowds out the healthy blood cells that can fight infection. The study encompassed cases of patients with CLL with > 5000 lymphocytes in absolute value, present at the cytological examination of the blood smear, from the peripheral blood, and CLL with less than 10% prolymphocytes based on the peripheral blood smears, which were May-Grunwald Giemsa stained **[Photo 3].**

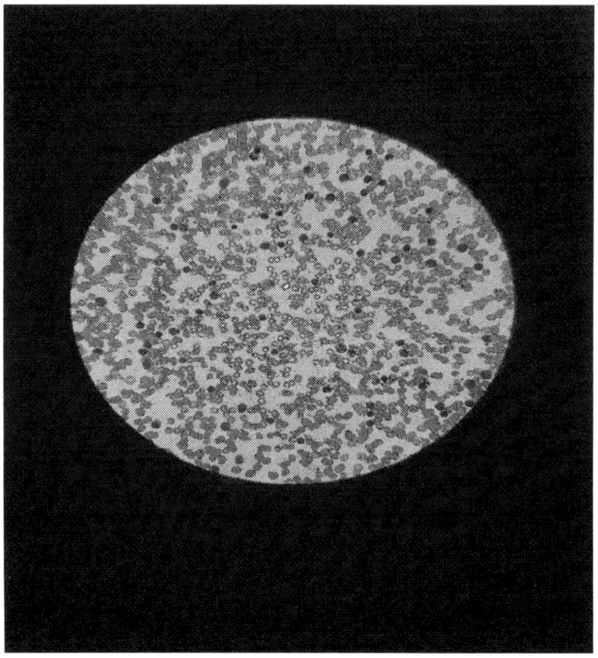

Photo 3. *Lymphocytes with Gumprecht nuclear shadows in CLL on a peripheral smear, colored with May-Grunwald Giemsa staining.*

The diagnosis of CLL-B was confirmed by immune phenotyping. All samples that entered the study were lymphocytes with positive $CD19^{+}$, $CD20^{+-}$, $CD5^{+}$ and $CD23^{+}$ cell receptors. The CD38+ receptor was considered positive if the distinct lymphocytes of the population showed a higher intensity of staining than the granulocytes in the sample and were associated with the presence of the protein ZAP-70, with pure prognostics of treatment in LLC with May-Grunwald Giemsa staining.

Participant type

Patients diagnosed with CLL-B were hospitalized in the Hematology departments of the Oncology Institute between November 2016 and September 2019. Age group: Senior. Gender: Both. Target number of participants: 20. Recruitment start date: 01/11/2016. Recruitment end date: 01/09/2019. Location: Romania.

Materials and method used with ELISA

The study used the sandwich ELISA quantitative method for direct detection of the p53 isoform protein, the product of the gene p53. Specificity: human p53 protein (aa20-25). Format: purified product. Monoclonal antibody clone: Isotype DO-1: IgG2a. The antibody is suitable for the following techniques: ICC/IF and ELISA. The research PAb 240 antibodies bind specifically to denatured p53 protein, in conformity with Wang et al. (2001). Compatible sample types: cell culture supernatants, plasma and serum. Solid support: 96-well microplate.

The principle of the ELISA method

This analysis is based on the sandwich ELISA principle. Each well of the microtiter plate was pre-coated with a specific target capture antibody. Standards or samples are added to the wells and the target antigen, in this case the p53 protein, binds to the capture antibody.

Protocol of ELISA work

1. Prepare all reagents, samples and standards as instructed in the manual.
2. Add 100 μl of standard or sample to each well.
3. Incubate 2.5 h at room temperature (RT), or past night, at 4°C.
4. Add 100 μl of prepared biotin antibody to each well.
5. Incubate 1 h at RT.
6. Add 100 μl of prepared streptavidin solution to each well.
7. Incubate for 45 min at RT.
8. Add 100 μl of TMB one-step substrate reagent to each well.
9. Incubate for 30 min at RT.
10. Add 50 μl of stop solution to each well.
11. Read at 450 nm immediately on ELISA analyzer.

After analyzing the 85 LLC samples, in different stages of disease evolution, starting with stage zero, and up to stage IV, only 20 patients were selected to be investigated for the detection of the p53 protein isoforms responsible for resistance to oncological treatments of the disease with rituximab, cyclophosphamide, doxorubicin hydrochloride, vincristine sulfate (Oncovin), and prednisone (R-CHOP), after 2 cycles of relapses, representing a group of 16 men and 4 women aged 39-85 years.

Male results: Protein concentration in p53/µg/dL: 20, 15, 18, 40, 10, 12, 14, 60, 30, 10, 13, 15, 5, 10, 15, and 12.

Female results: Protein concentration in p53/µg/dL: 140, 30, 13, and 10. Normal values of normal cell lines on equipment: ELISA = 0.25-0.5 µg/dL, or 2.5- 5ng/ml.

Statistical Interpretations:

The concentration of the p53 isoform protein representing the p53 mutant gene in 17 cases, after excluding the 3 outlier cases present in the study, was calculated at the mean value of 14.8 µg/dL, with standard deviation, STDEV = 6.46, and CV = 0.4%. The probability index (NORMDIST), "p", was calculated at the value of $p = 0.079$. Multivariate (MA) analyses were performed for OS on any significant variables at the level of $p < 0.20$ at univariate analysis, with the gradual elimination of insignificant variables.

The presence of protein isoforms in the group was calculated as at 17%, and the un-favorable evolution and transformation of Chronic Lymphocytic Leukemia in the stages studied in Diffuse Large Lymphoma were calculated to be at 3.5%, as in the meta-analyzes from the specialized literature. Diffuse Large lymphoma is considered a rare disease. The comparison between the categorical variables and the numerical variables was performed using the exact Fisher test.

Fisher test

Number of patients in the study group = 85; Eligible 20 a + b = 3 + 17; Ineligible 65. c + d = 5 + 60. Total a + c b + d a + b + c + d (= n) = 85; p = {a + b} \ {a} x {c + d} \ {c}: [{n} \ {a + c}] = 20 / 3x65 / 5 / 8,125 = 10.2 = 1.1 log in base 10. The result indicated that all statements were true: {\ frac {(a + b)! (c + d)! (a + c)! (b + d)!} [a! b! c! d! n!], in the calculations made by the statistical software where, (n / k) was the binomial coefficient and the symbol "!" showed that factorial **[Graphic 1].**

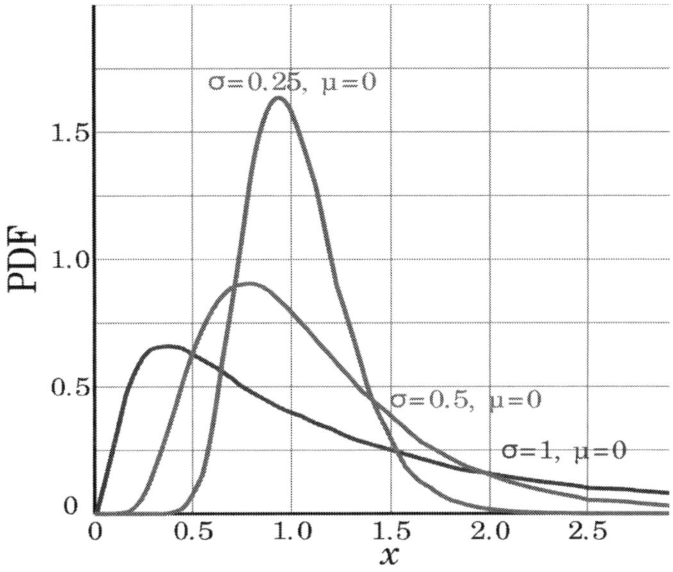

Graphic 1. *All expressions under the sign of the logarithm are indicated as positive which is under the sign of the logarithm and as a basis, the remaining expression below the logarithm is the frequency of p53 isoform protein in the cohort of the study was evaluated with ANOVA test.*

Calculation of the frequency of isomorph p53 protein by ANOVA tests was positive, in value 1:1, indicating a linear relationship between the 2 variables **[Table 1]**.

The interval of concentration **(i)**	**The middle classes (m)**	**Frequency of concentration (f)**
30-32	31	1
27-29	28	0
24-26	25	0
21-23	22	0
18-20	19	2
15-17	16	2
12-14	13	8
9-11	10	4

6-8	7	0
3-5	4	1 17%

Table 1. *Calculation of frequency of isomorph p53 protein by ANOVA test positive.*

Very high pathological values in the 3 cases of p53 were calculated in the 2 men at the values of 60 μg/dL and 40 μg/dL, and in the case of the female it was calculated in the amount of 140 μg/dL - the frequency chronic lymphocytic leukemia with transformation into Diffuse Large Lymphoma (DLL) **[Graphic 2]**.

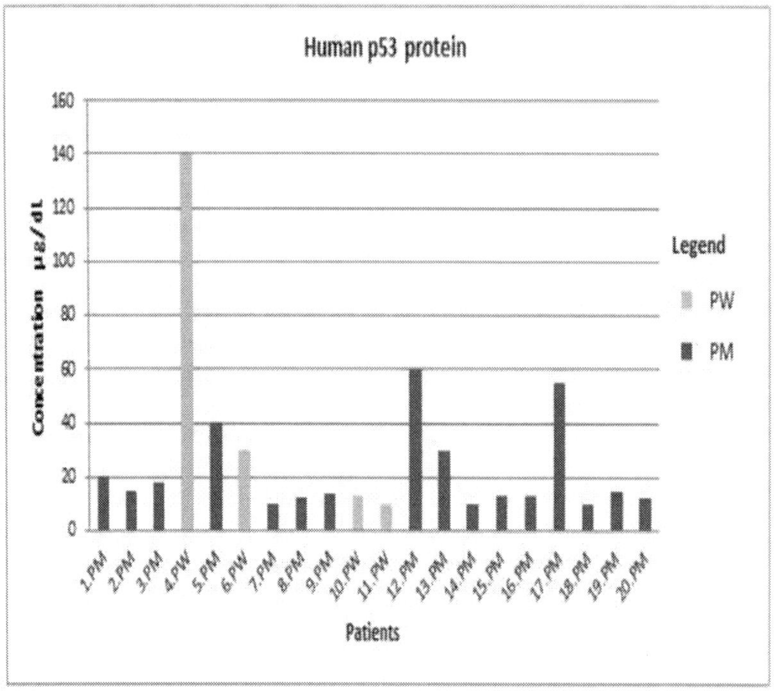

Graphic 2. *Very high pathological values in the 3 cases of p53 were calculated in the 2 men with the values of 60 μg / dl and 40 μg / dl, and in the case of the female it was calculated in the amount of 140 μg / dl - the frequency of Chronic Lymphocytic Leukemia with transformation into Diffuse Large Lymphoma (DLL).*

The pathological values in the 3 cases of highly elevated p53, reflecting the concentration of p53 protein mutant, were calculated in 2 men in the amount of 60 μg/dl and 50 μg/dl, and in the female case, it was calculated in the amount of 140 μg/dl - the frequency of chronic lymphocytic leukemia being higher in men than in women in a 2:1 ratio, in accordance with the data from the specialized literature. The overall frequency of positivity of the p53 protein in the increased number of CLL cases studied was 15% (3 out of 20 cases). The expression of the high-concentration p53 proteins in stages 2 and 3 of the disease was associated with a significantly weaker response to chemotherapy (p = 0.034) **(Table 2).**

CLL – Age patients	**CLL stage I/II, (n=17 patients)** **P53 protein concentration in reactive limfocytes B**	**CLL stage III/IV (n = 3 patients]** **Percentage of p53 izoform proteins**
The age of patients with LLC, ranging from 39 to 85 years.	*The average p53 protein concentration in CLL, 16.76 μg / dL*	*P-5 izofoorm proteins with elevated values was present in 15% (3 of 20 cases* *2 Men = 50μg / dL and 60μg / dL, respectively* *1 Female = 140μg / dL*
Hematological parameters in peripheral blood	***Mean values of haemogram:*** *No. Leukocytes = 35-50 x 10^3/μL* *Hb = 11.8g / dL;* *Platelet = 140 x 10^3/μL* *Lymphocytes = 65-80%*	*No. Leucocytes = 250-500 x 10^3/μL* *Hb = 8.6g / dL* *Thrombocytosis = 45x10^3/μL* *Limphocytes = 85-90%*

Table 2. *Expression of hemogram parameters and p53 protein concentration in different stages of CLL-B.*

The main stimulus for the malignant tumor is hypoxia from the predominant anaerobe metabolism. The aim of this work was to emphasise the levels of energy in cells, in types of malignant diseases, due to the disregulation of genes, using the bioluminescence method by running a Luminometer Analyzer with bio-luminescence LKB.

The measurement of ATP was made by the standard principle of bioluminescence on the automatic analyzer LKB, using ATP monitoring reagent, ATP standard, (106 Mol ATP/ ml), buffer solution Tris-EDTA, and TCA-EDTA 1 lyse. All results were statistically analyzed by the Excel program. The principle of the reaction was: ATP + luciferin + O2----> oxyluciferin + AMP + PPi + CO2 + Light. The reaction was catalyzed by the enzyme luciferase from the ATP research kit.

Working technique

Lymphocytes were obtained from the peripheral blood of healthy individuals and hospitalized patients with selected diseases from the oncology departments. The plasma was collected from patient samples using vacutainers with EDTA or heparin as an anticoagulant by centrifuging the samples for 15 minutes at 4500 rpm (280 G).

After the blood sample has been centrifuged and its plasma has been separated from red blood cells, the plasma is fractionated into four distinct fractions and placed on a layer of white blood cells (lymphocytes). With a pipet, a quantity of 100μL is extracted from the lymphocyte ring. The extracted lymphocytes are introduced into 25 ml cuvettes with a 3 ml Wash Buffer medium for the washing of the lymphocytes **[Photo 4]**.

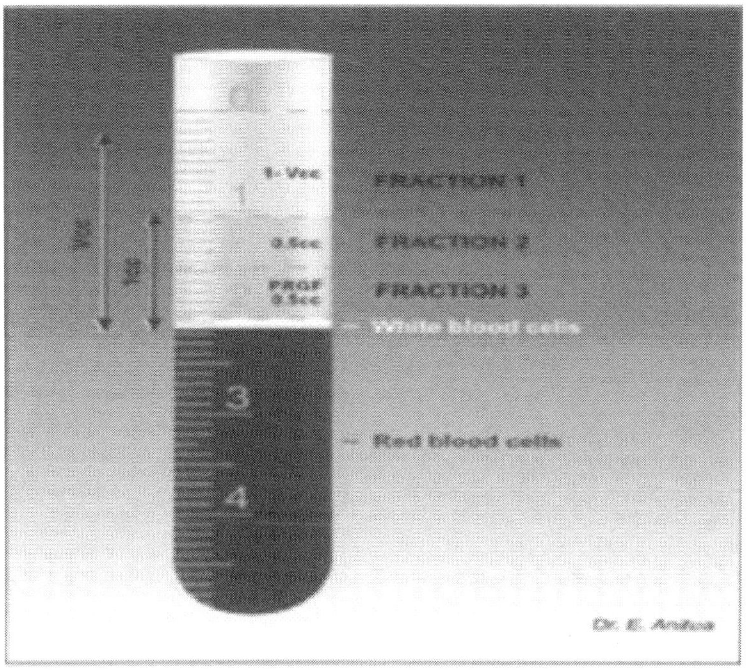

Photo 4. *Lymphocyte layer required for drop of 100μL in 3 ml Wash Buffer medium.*

Washing is done 3 times, once after 10 minutes, at 1500 revolutions/ minute, and again after 20 minutes and 30 minutes at 1000 revolutions/ minute. The lysis of washed lymphocytes is done with a microwave device - the Mini Wave Smart Laboratory microwave. Activated B lymphocytes were defined as $CD19^{+}$, $CD20^{+-}$, and $CD21^{+}$ cells, and $CD23^{+}$ cells and/or a $CD138^{+}$ surface marker. Activated T cells were identified with CD7, CD5, CD3, CD2, CD4, CD8 and CD45RO. The phenotype suggestive of B-CLL or monoclonal B cell lymphocytes was approximately equal numbers of $CD5^{+}$ and $CD19^{+}$ lymphocytes.

Concentration analysis of ATP in mean values, ($\bar{x}$), with Standard Deviation, (SD) : -ATP in normal T Cells,$\bar{x}$ = 1.39 μM ATP, (SD = 0.41); ATP in normal B cells, $\bar{x}$=0.35 μM ATP, (SD = 0.42); ATP in T cells of malignant diseases, $\bar{x}$= 0.17 μM ATP, (SD = 0.46);

-ATP in B cells of malignant disease=, $\bar{x}$=3.06

μM ATP, (SD = 0.45); -ATP in B cells of Leukemia, $\bar{x}$= 4.33 μM ATP, (SD =1.5); - ATP in T cells of

Leukemia, $\bar{x}$= 0.09 μM ATP, (SD=1.7) **[Table 3]**.

Normal conc. ATP in T Cells	**Normal conc. ATP in B cells**	**Conc. ATP in T cells, malignant diseases**	**Conc. ATP in B cells, malignant disease**	**Conc. ATP in B cells, CLL**
$\bar{x}$ = 1.39	**$\bar{x}$ =0.35**	**$\bar{x}$ == 3.06**	**$\bar{x}$ = 0.17**	**$\bar{x}$ =4.33**
SD = 0.41	**SD = 0.42**	**SD = 0.46**	**SD = 0.45**	**SD =1.5**

***Table** 3. The concentration of micromoles, μM ATP, in malignant diseases were determined with Bio-luminescence.*

Regarding the study of malignant diseases, it was noted the increased measured values: 4.30 - 4.55 μM ATP for chronic lymphatic leukemia and respectively 3.20 - 3.65 micromoles ATP for neoplasms or with bone metastases. Ongoing research has highlighted the fact that the direction of the synthesis reactions ADP + Pi>-<ATP + H2O can be controlled in "vitro", depending on the concentration of system factors, and pH and ionic strength, making it likely that there is a functional energy control by creating weak bonds between the chemical complexes of the newly formed biologically active substances.

Deletion in the short arm of chromosome 17 is associated with rapid progression, short remission, and decreased overall survival in chronic CLL. 17p13 deletions are associated with the loss of function of the tumor suppressor gene p53 and deletions of the bands 11q22-q23 and are associated with extensive lymph nodes, more aggressive diseases and shorter survival times.

The structural and biochemical changes in malignant cells will influence the development of characteristic bioenergetic processes that at their level will condition the expression of cellular oncogenes. The p53 protein is an oligomeric transcription factor of 393 residues and is organized into 5 structural and functional regions **[Scheme 16]**.

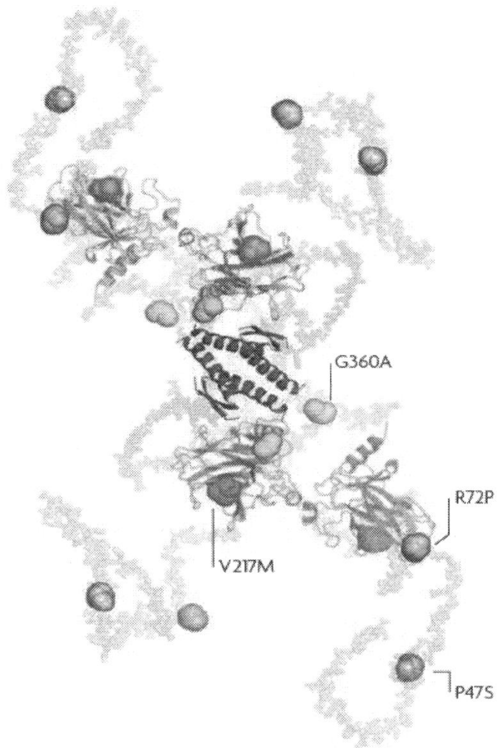

Fig 3. Three-dimensional structure of p53 in its tetrameric form. The four coding sequence polymorphisms discussed in the text are indicated. Glycine 360 (polymorphic variant is alanine) is displayed in yellow; valine 217 (polymorphic variant is methionine) is displayed in green; arginine 72 (polymorphic variant is proline) is displayed in pink; proline 47 (polymorphic variant is serine) is shown in purple. Figure is modified, with permission, from [160] © (2007) National Academy of Sciences, USA. Figure courtesy of A. Fersht, University of Cambridge, UK

Scheme 16. *Three-dimensional structure of isomorph p53 protein in tetrameric form.*

Restoring p53 function can induce cell death and apoptosis consistent with this role, the activity of the p53 protein is compromised to a greater or lesser extent in all types of cancer, either by mutation of the p53 gene or changes in the modulating state of the p53 protein. Theoretical studies have suggested that elevated concentrations of adenosine triphosphate (ATP) in malignant cells.

B cells' lymphocytes from CLL impair the p53 gene to induce apoptosis of cells. Apoptosis is regulated by a cascade of proteins called caspases that are present in all cell forms as pro-forms. After cleavage, caspases become active and initiate pathways leading to apoptosis.

The death resistance of hypoxia takes place on at least two levels, within mitochondria and the cytosol. Accumulation of Bax in the mitochondria causes the release of cytochrome C in cytosol, which is strongly reduced under hypoxic environment conditions **[Scheme 17]**.

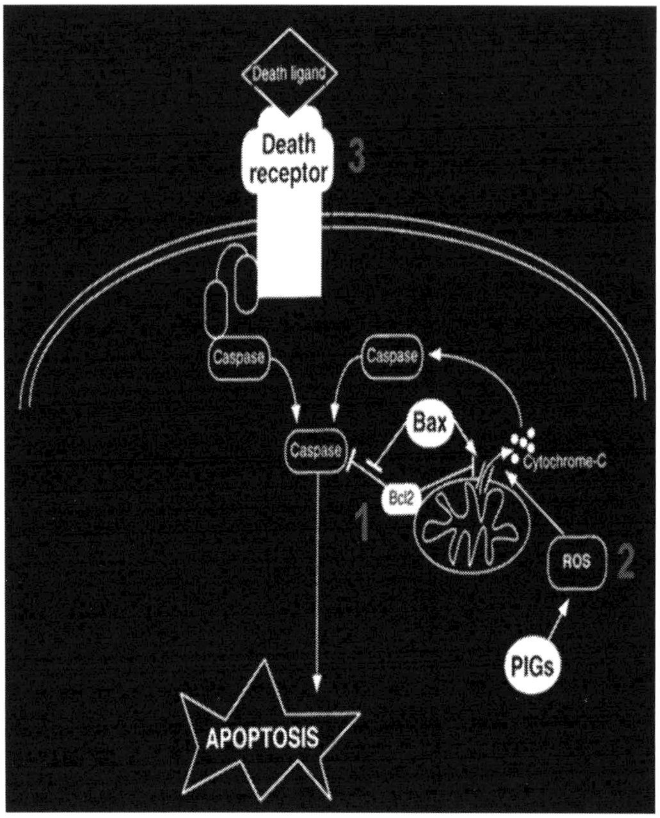

Scheme 17. *The pro-apoptotic proteins Bax, Bad, Bak and Bid induce programmed cell death.*

Apoptosis, performed within the extrinsic pathway, is initiated by death ligands, such as the Fas ligand or the tumor necrosis factor (TNF), and ultimately leads to the activation of caspases 8 and 3 and the activation of the apoptosis protein activating factor (APAF).

All hematopoietic cells, especially lymphoid cells, express Bcl-2, mainly in the nucleus and endoplasm reticulum. The oncogene Bcl-2 may be a general suppressor of genes that directly regulate apoptosis.

Each cell needs the energy to maintain its vital structure, for the synthesis of reserve substances and the exercise various specific cellular functions and various movement processes; plasma movement, nuclear division and cell synthesis. As in normal cells, the energy-generating substances are primarily carbohydrates and then fats and proteins.

The phenomenon is particularly prevalent in aggressive malignancies, most of which are also hypoxic. The hypoxia induces a stochastic imbalance between the numbers of reduced mitochondrial species and the available oxygen, resulting in increased reactive oxygen species (ROS) whose toxicity can lead to apoptotic cell death [**Scheme 18**].

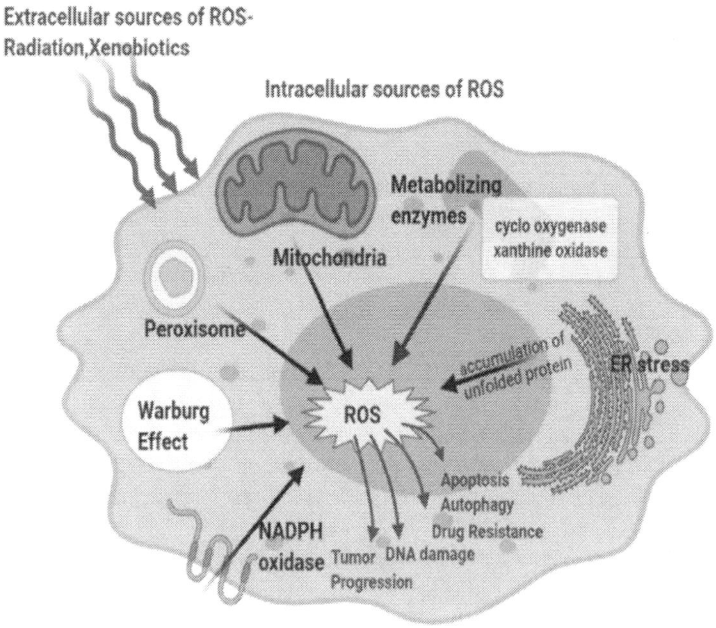

***Scheme 18**. Free oxygen radicals (ROS) of metabolism from various dietary factors may be a major cause of carcinogenesis.*

The alteration of the physiological processes of oxidative phosphorylation, the decrease of the respiratory level and the increase of anaerobic glycolysis in the processes of malignancy are accompanied by the increase in the intensity of cellular bioluminescence. Bioluminescence thus appears as an intrinsic component of metabolism and not as the product of specialized photogenic organs.

All living cells contain the macro-energetic molecule, adenosine triphosphate (ATP), which is present in a fairly constant amount for each cell type and is rapidly lost after cell death. It was found that the electrical charge of the cell membrane varies relatively little around the value of 0.85eV, in resting conditions. If the energy charge decreases below 0.85 eV, (ATP (<1.75 μM in a cell), the ATP-generating sequences are accelerated by the response of the respective regulatory components, according to the metabolic needs.

ATP is the energy base of DNA-polymerase and DNA-ligase enzymes, enzymes that repair altered DNA nucleotides under the influence of cellular aggression induced by intracellular (free radicals) or extracellular (physical, chemical and biological) toxic factors. The p53 protein, the product of the p53 gene, has additional roles, which may overlap with its suppressive capacity, in tumor processes, including the response to DNA damage, ageing metabolism and stem cell differentiation.

Following this information, we can say that a certain type of malignant disease is due to the dysregulation of genes at a certain time and way of life. The structural and biochemical changes in malignant cells will influence the development of characteristic bioenergetic processes that at their level will condition the expression of cellular oncogenes to the detriment of anti-oncogenes that are expressed up to now, normally in an aerobic metabolism **[Scheme 19]**.

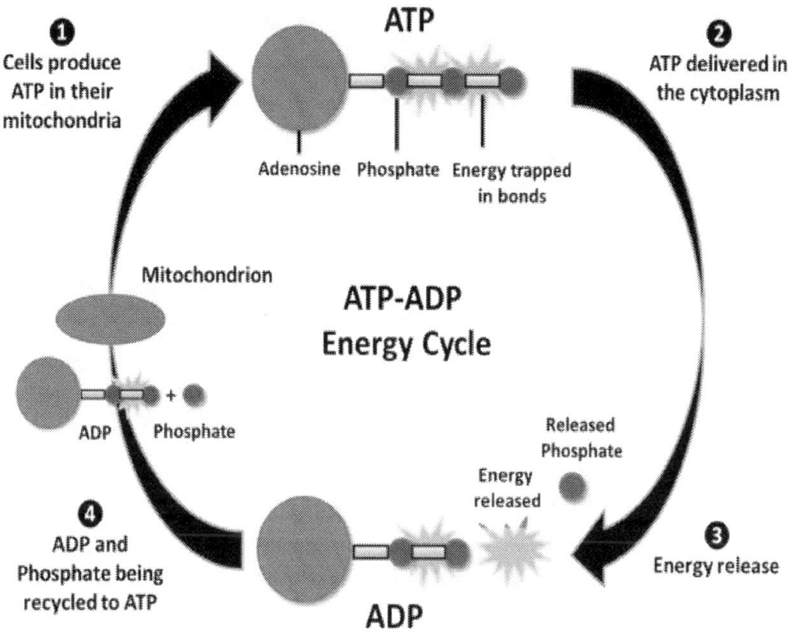

***Scheme 19**. Cytoplasmic phosphorylation pathways in a cell with predominantly anaerobic metabolism.*

Otto Warburg observed that many cancers lose their capacity for mitochondrial respiration, limiting ATP production to anaerobic glycolytic pathways. Ongoing research has highlighted the fact that the direction of the synthesis reactions ADP + Pi>---<ATP + H2O can be reproducibly controlled in vitro, depending on the concentration of system factors, pH and ionic strength, making it likely that a functional energy control exists by creating weak bonds between the chemical complexes of the newly formed biologically active substances **[Scheme 20]**.

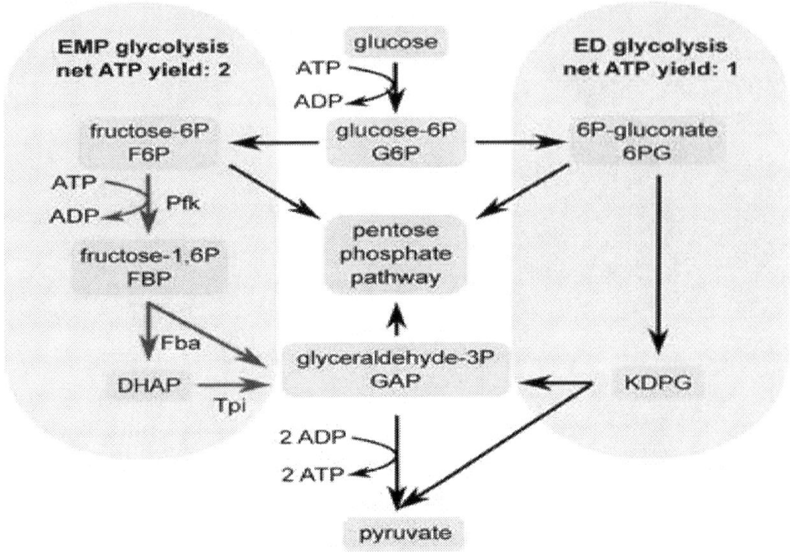

Scheme 20*. Pathways of ATP production in anaerobic glycolytic pathways.*

The mechanism involves inhibition of glycolytic ATP production via a Randle-like cycle while increased uncoupling renders cancers unable to produce compensatory ATP from respiration generation in the presence of intact tricarboxylic acid (TCA) enzyme.

One mitochondrial adaptation to increased ROS is the over-expression of the uncoupling protein 2 (UCP2) that has been reported in multiple human cancer cell lines.

Increased UCP2 expression was also associated with reduced ATP production in malignant oxyphilic mouse leukemia and human lymphoma cell lines. Hypoxia reduces the ability of cells to maintain their energy levels because less ATP is obtained from glycolysis than from oxidative phosphorylation. Cells adapt to hypoxia by activating the expression of mutant genes in glycolysis. Severe hypoxia causes a high mutation rate, resulting in point mutations that may be explained by reduced DNA mismatch repairing activity.

The most direct induction of apoptosis caused by hypoxia is determined by the inhibition of the electron carrier chain from the inner membrane of the mitochondria. The lack of oxygen inhibits the transport of protons and thereby causes a decrease in membrane potential.

Cell survival under conditions of mild hypoxia is mediated by phosphoinositide-3 kinase (PIK3) using severe hypoxia or anoxia, and then cells initiate a cascade of events that lead to apoptosis. After DNA damage, a very important regulator of apoptosis is the p53 protein. This tumor suppressor gene has mutations in over 60% of human tumors and acts as a suppressor of cell division.

The genetic changes can be identified by fluorescent probes to chromosomal using a technique referred to as fluorescent in situ hybridization (FISH).

Chromosomal evaluation using FISH can identify certain chromosomal abnormalities of CLL that have prognostic significance. The deletion of the part of the short arm of chromosome 17 (del 17 q) which targets the cell cycle regulating protein p53 is particularly deleterious **[Photo 5]**.

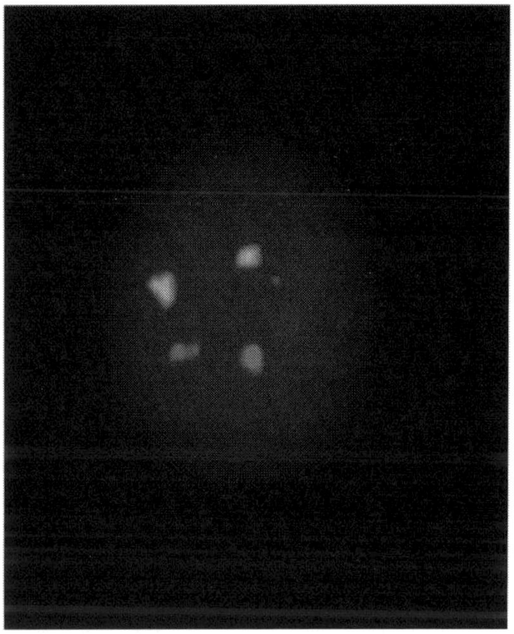

Photo 5. *The normal model of fluorescence signals of the p53 gene.*

This abnormality is found in 10% of patients with CCL and has a pure prognosis. Deletion of the long arm of chromosome 13(del 13q) is the most common genetic abnormality in CLL with roughly 50% of patients exhibiting this effect. These patients have the best prognosis and most will live many years without the need for therapy. Agents damaging DNA may increase the expression of p53 and its trans-activation activity, suggesting that p53 acts protect cells against the accumulation of mutants and their subsequent conversion to a malignant status [27].

Hypoxia itself can also prevent apoptosis by inducing the expression of the anti-apoptotic protein IAP-2. A typical response to the hypoxic environment, by hypoxia-inducible factor 1 for example, is the expression of insulin-independent GLUT triggered by HIF 1α ensuring maximum glucose uptake for glycolytic ATP generation.

p53 protein, in its normal form, acts in stopping the cell division whenever damage to a cell's DNA is detected, thus giving the cells the possibility of repairing DNA before the errors would duplicate and be passed on to the daughter cells. Over-expression of normal p53 protein can result either in G1 arrest, mediated by p21 protein or in the induction of apoptosis.

Survivin is a 16.5 KD protein and the smallest inhibitor of apoptosis so far identified. It is expressed in the G2/M phase of the cell cycle in a cycle-regulated manner. At the beginning of mitosis surviving associates with the microtubules of the mitotic spindle in a specific and saturable reaction that is regulated by microtubule dynamics **[Scheme 21]**.

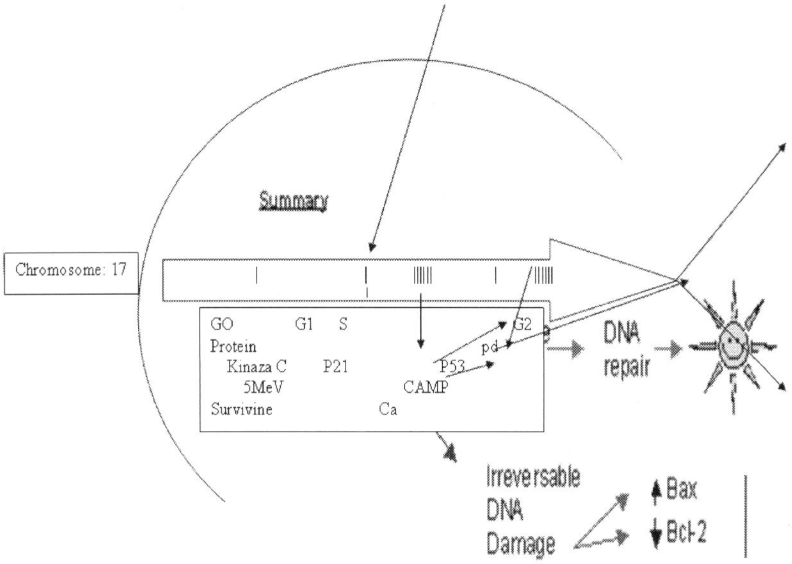

Fig5. Protein p21 is a cyclin-dependent kinase inhibitor that prevents cell cycle progression at G1. It is induced by wild-type p53 following DNA damage, and p21-null cells are deficient in DNA damage-induced, p53-mediated G1 arrest.

Scheme 21. *The p53 protein is becoming an attractive therapeutic target for anticancer drug discovery.*

The over-expression of survivin in cancer may overcome this apoptotic checkpoint and favor the aberrant progression of transformed cells by mitosis. In analyzing this process of initiating carcinogenesis, the Desing Human Survivin Titer Zyme Enzyme Immunometric Assay (EIA) is acceptable to support the analysis of the plasma sample. Severe and prolonged hypoxia may initiate apoptosis whereas under acute and mild hypoxia cells adapt to this environmental stress and will survive [28].

Antibodies to human p53 have been detected in patients with cancer. These antibodies are highly specific for malignant diseases and are rarely detected in healthy donors or patients having benign diseases. This immune response is correlated with the presence of a p53 gene mutation, leading to the accumulation of an ineffective p53 protein in tumor cells with either tridimensional structure [29].

Biological targets related to CLL starting from Bruton tyrosine kinase

BTK was placed in the signal transduction pathway downstream of the B-cell receptor (BCR). This receptor is expressed on the B-cell surface and has the unique capacity to specifically recognize antigens due to hypervariable regions present in the immunoglobulin heavy (IGH) and light (IGL) chains that together form the BCR. BTK is also involved in many other signaling pathways in B cells, including chemokine receptor, Toll-like receptor (TLR) and Fc receptor signaling. Expression of BTK is not restricted to B cells, as also cells of the myeloid lineage express BTK [29].

In these cells, BTK also acts downstream of TLRs, for example in the FcεR in mast cells and the Fcy RI in macrophages. In addition, BTK is involved in various other pathways, including the receptor activator of nuclear factor-Kb, in osteoclasts collagen and in CD32 signaling in platelets and the NLRP3 inflammasome in macrophages and neutrophils.

Since myeloid cells are important components of the tumor microenvironment and particularly tumor-associated macrophages contribute to cancer progression, there is currently considerable interest in BTK inhibition as an anti-cancer therapy not only in B cell leukemias but also in other hematological malignancies and solid tumors.

BTK is essentially cytoplasmic and is only transiently recruited to the membrane through the interaction of its PH domain with phosphatidylinositol-3,4,5-triphosphate (PIP3), which is generated by phosphatidylinositol-3 kinase (PI3K). BTK activation occurs in two steps upon its recruitment to the cell membrane. BTK-deficient B cells enter early G1, but not the S phase of the cell cycle, because they fail to induce cyclin D2 expression [30].

ATP-dependent Chromatin Remodeling Chromatin remodeling is the enzyme-assisted movement of nucleosomes on DNA. Chromatin is a condensed and often inaccessible structure where genomic DNA is packaged through histone and non-histone proteins. When DNA damage occurs, efficient and accurate repair of DNA damage ensures genome stability and prevents damage development which could lead to cancer or cell death.

Activating DNA damage response (DDR) enables the cells to utilize post-translation histone modifications and ATP-dependent chromatin remodeling to modulate chromatin structure and increase the accessibility of the repair machinery to lesions embedded in chromatin, Chromatin remodeling utilizes the energy of ATP to disrupt nucleosome DNA contacts, move nucleosomes along DNA, and remove or exchange nucleosomes such that DNA repair can be accomplished.

Via ATP hydrolysis, the chromatin structure of a number of large multiprotein complexes (200 kDa2- MDa) can be enzymatically modulated. Several chromatin remodeling complexes are involved in the process: switch/sucrose non-fermentable (SWI/SNF) family containing either the Brahma (BRM) or Brahma-related gene 1 (BRG1). ATPase which slide and eject nucleosomes, imitation switch (ISWI) complexes containing SNF2H or SNF2L ATPase and mediate nucleosome sliding and histone displacement, inositol requiring 80 (INO80) chromatin remodeling factors containing INO80 ATPase or related SWR1-like factors such as the p400 ATPase which features long insertion in the middle of the conserved ATPase domain, and chromodomain helicase DNA binding protein (CHD) family members containing two tandemly arranged chromodomains (CDs) on the N-terminus of their ATPase which is involved in binding methylated histone tails as well as DNA and can slide and eject histones and have both activator and inhibitory roles in transcription regulation.

An ATPase which is capable of DNA translocation moves nucleosomes such that transcription factors can access DNA. Luijsterburg and van Attikum recently linked chromatin and the DNA damage response with cancer. Hargreaves and Crabtree reviewed the genetics, genomics and mechanisms of ATP-dependent chromatin remodeling. While many cancer cells have defects in one or more aspects of the DDR, such cells may be more vulnerable to cancer therapies that aim at targeting the tumor-related DDR defects.

1.1. A Janus-activated kinase (JAK)–STAT pathway leading to impaired transcriptional activity

Janus Kinase (JAK) protein tyrosine kinases are recognized as an integral component of the cytokine receptor subunits, and enzyme activation as the initiating step in a signaling cascade required for embryonic development, tissue growth, haemopoietic development and differentiation, innate and adaptive immunity and the inflammatory response. Given that JAK-STAT signaling is the universal and essential intracellular pathway for cytokine action, a number of regulatory mechanisms have evolved to control the magnitude and duration of signaling.

This allows fine-tuning of cytokine-mediated cellular effects and prevents the inappropriate activity often associated with disease development. There are three major mechanisms for negative regulation: receptor internalization, de-phosphorylation by phosphor-tyrosine phosphatases (PTPs), and direct inhibition by protein inhibitors of STATs (PIAS) and suppressor of cytokine signaling (SOCS) proteins.

While loss-of-function mutations in JAK3 and TYK2 are associated with immunodeficiency, the majority of naturally occurring mutations in JAK2, including chromosomal translocation, point mutations, insertions and deletions, are gain-of-function mutations and are associated with acute leukemia or myeloproliferative disorders (MPDs). The chromosomal translocations of JAK2 loci (for example, TEL-JAK2, PCM1-JAK2, BCR-JAK2 and PAX-JAK2) lead to the development of both myeloid and lymphoid hematological malignancies. These translocations result in the fusion of the JAK2 catalytic kinase (JH1) domain with multimerization subunits of partner proteins leading to constitutive tyrosine kinase activity and transformation.

Given the involvement of the V617F mutation in MPNs, most of the therapeutic development has focused on identifying low molecular mass ATP-competitive JAK2 inhibitors, reviewed in (296,297). TG101348 and Ruxolitinib (or INCB018424) (298,299) are two of many selective and potent JAK inhibitors currently in use in phase II and III MPN clinical trials, respectively. Ruxolitinib has a strong inhibitory effect on both JAK1 and JAK2, while TG101348 has greater selectivity for JAK2.

Both of these inhibitors effectively attenuate downstream signaling, including phosphorylation of STAT and ERK1/2, and induce apoptosis in vitro. MPN patients treated with these inhibitors demonstrated a dramatic improvement in constitutional symptoms and a reduction in spleen size. Although unpleasant side effects and a less satisfactory improvement in cytopenia, marrow fibrosis and JAKV617F burden were also observed, inhibition of JAK2 has already proven to be an excellent target for therapeutic intervention in MPNs.

AK-STAT signaling is able to interconnect with other cell-signaling pathways, such as the PI3K/AKT/mTOR pathway. When JAKs are activated and phosphorylate tyrosine residues on receptors, proteins with

SH2 domains (such as STATs) are able to bind to the phosphor-tyrosine, and the proteins can carry out their function. Like STATs, the PI3K protein also has an SH2 domain, and therefore it is also able to bind to phosphorylated receptors. As a result, activating the JAK-STAT pathway can also activate PI3K/AKT/mTOR signaling.

JAK-STAT signaling can also integrate with the MAPK/ERK pathway. Firstly, a protein important for MAPK/ERK signaling, called Grb2, has an SH2 domain, and therefore it can bind to receptors phosphorylated by JAKs (in a similar way to PI3K). Grb2 then functions to allow the MAPK/ERK pathway to progress. Secondly, a protein activated by the MAPK/ERK pathway, called MAPK (mitogen-activated protein kinase), can phosphorylate STATs, which can increase gene transcription by STATs.

However, although MAPK can increase transcription induced by STATs, one study indicates that phosphorylation of STAT3 by MAPK can reduce STAT3 activity. One example of JAK-STAT signaling integrating with other pathways is Interleukin-2 (IL-2) receptor signaling in T cells. IL-2 receptors have γ (gamma) chains, which are associated with JAK3, which then phosphorylates key tyrosine on the tail of the receptor.

Phosphorylation then recruits an adaptor protein called Shc, which activates the MAPK/ERK pathway, and this facilitates gene regulation by STAT5. Given that many JAKs are associated with cytokine receptors, the JAK-STAT signaling pathway plays a major role in cytokine receptor signaling. Since cytokines are substances produced by immune cells that can alter the activity of neighbor cells, the effects of JAK-STAT signaling are often more highly seen in cells of the immune system.

For example, JAK3 activation in response to IL-2 is vital for lymphocyte development and function. Also, one study indicates that JAK1 is needed to carry out signaling for receptors of the cytokines IFNγ, IL-2, IL-4 and IL-10.

The JAK-STAT pathway in cytokine receptor signaling can activate STATs, which can bind to DNA and allow the transcription of genes involved in immune cell division, survival, activation and recruitment. For example, STAT1 can enable the transcription of genes which inhibit cell division and stimulate inflammation.

Also, STAT4 is able to activate NK cells (natural killer cells), and STAT5 can drive the formation of white blood cells.

In response to cytokines, such as IL-4, JAK-STAT signaling is also able to stimulate STAT6, which can promote B-cell proliferation, immune cell survival, and the production of an antibody called IgE. AK-STAT signaling plays an important role in animal development. In some flies with faulty JAK genes, too much blood cell division can occur, potentially resulting in leukemia. JAK-STAT signaling has also been associated with excessive white blood cell division in humans and mice.

Mutations in JAK2 can lead to leukemia and lymphoma. Specifically, mutations in exons 12, 13, 14 and 15 of the JAK2 gene are proposed to be a risk factor in developing lymphoma or leukemia. Additionally, mutated STAT3 and STAT5 can increase JAK-STAT signaling in NK and T cells, which promotes very high proliferation of these cells, and increases the likelihood of developing leukemia.

Also, a JAK-STAT signaling pathway mediated by erythropoietin (EPO), which usually allows the development of red blood cells, may be altered in patients with leukemia.

1.2. Phosphoinositide 3-kinase (PI3K)/AKT pathway resulting in increased apoptosis

The phosphoinositide 3-kinase (PI3K) pathway is frequently activated in human cancer and represents an attractive target for therapies based on small molecule inhibitors. PI3K isoforms play an essential role in the signal transduction events activated by cell surface receptors including receptor tyrosine kinases (RTKs) and G-protein-coupled receptors (GPCRs). There are eight known PI3K isoforms in humans, which have been subdivided into three classes (I-III).

Therefore, PI3Ks show considerable diversity and it remains unclear which kinases in this family should be targeted in cancer.

The class IA of PI3K comprises the p110α, p110β and p110δ isoforms, which associate with activated RTKs. In human cancer, recent reports have described activating mutations in the PIK3CA gene encoding p110α, and inactivating mutations in the phosphatase and tensin homologue (PTEN) gene, a tumor suppressor and antagonist of the PI3K pathway.

The PIK3CA mutations described in cancer constitutively activate p110α and, when expressed in cells drive oncogenic transformation.

Moreover, these mutations cause the constitutive activation of downstream signaling molecules such as Akt/protein kinase B (PKB), mammalian target of rapamycin (mTOR) and ribosomal protein S6 kinase (S6K) that is commonly observed in cancer cells. In addition to p110α, the other isoforms of the PI3K family may also play a role in human cancer, although their individual functions remain to be precisely identified.

In this review, we will discuss the evidence implicating individual PI3K isoforms in human cancer and their potential as drug targets in this context. In addition, PI3K was reported to be involved in the activation of several other protein kinases, including c-Jun N-terminal kinase (JNK) by EGF. Bruton's tyrosine kinase (Btk), which has a PH domain that can bind PIP3.

RNA interference (RNAi) as well as conditional knock-out of ILK had no effect on phosphorylation of Akt on Thr-308 but resulted in almost complete inhibition of phosphorylation on Ser-473 and significant inhibition of Akt activity, accompanied by significant stimulation of apoptosis. In addition, Raf-1 activation by Ras was shown to be achieved through a combination of both physical interaction and indirect mechanisms involving the activation of PI3K as a second Ras effector, which directs p21-activated kinase (PAK)-mediated regulatory phosphorylation of Raf-1.

Phosphorylation of Raf-1 on Ser338 through PI3K and Pak was also shown to provide a co-stimulatory signal which together with Ras leads to strong activation of Raf-1 kinase activity by integrins. All cycle progression is a tightly controlled process. To initiate cell division, mitogens trigger a number of early signals that promote the G(0)-G, (1) transition by inducing cell growth and the activation of G(1) cyclins.

Activation of cyclin E/cdk2 (cyclin-dependent kinase 2) at the end of G, (1) is then required to trigger DNA synthesis (S phase entry). Among the early signals induced by mitogens, activation of PI3K appears essential to induce cell cycle entry, as it regulates cell growth signaling pathways (see previous section), which in turn determines the rate of cell cycle progression.

The mechanisms by which PI3K and its downstream effector Akt regulate cell cycle entry are by inactivation of the FOXO transcription factors, which induce expression of quiescence genes such as that encoding p27 (Kip), p130 and cyclin G2. PI3K/FOXO then work as a complementary switch; when PI3K is active, FOXO transcription factors are inactive.

PTEN expression also led to the inhibition of Akt/ PKB. It was also shown that the PTEN protein induces a G1 block when reconstituted in PTEN-null cells. A PTEN mutant associated with Cowden's disease (PTEN; G129E) has protein phosphatase activity yet is defective in dephosphorylating inositol 1,3,4,5-tetrakisphosphate in vitro and fails to arrest cells in G1.

These data suggest a link between the induction of a cell-cycle block by PTEN and its ability to dephosphorylate, in vivo, PIP3. In a parallel study, PTEN impaired the activation of endogenous Akt in cells and inhibited the phosphorylation of 4E-BP1. In addition, PTEN/ MMAC1 repressed gene expression in a manner that is rescued by Akt but not PI3K.

The PTEN tumor suppressor protein inhibits PI3K/Akt signaling that promotes translocation of Mdm2 into the nucleus. When restricted to the cytoplasm, MDM2 is degraded. The ability of PTEN to inhibit the nuclear entry of MDM2 increases the cellular content and transactivation of the p53 tumor suppressor protein. Retroviral transduction of PTEN into PTEN null glioblastoma cells increases p53 activity and expression of p53 target genes and induces cell cycle arrest U87MG/PTEN glioblastoma cells were more sensitive than U87MG/PTEN null cells to death induced by etoposide, a chemotherapeutic agent that induces DNA damage.

These results established a direct connection between the activities of two major tumor suppressors and show that they act together to respond to stresses and malignancies. PTEN almost twenty years after the discovery of PI3K, this family of enzymes has become the focus of intensive research aimed at developing new drugs for a wide variety of diseases, including cancer, inflammation and allergy.

The discovery of the many genetic alterations affecting the PI3K/PTEN/Akt/mTOR pathway in human cancer has clearly placed some of the enzymes involved, including p110α, mTOR and Akt, on the top list of druggable targets in the field of oncology research. Moreover, the recognition that some of the molecular alterations, such as increased Akt phosphorylation or PTEN mutations correlate with clinical parameters, such as outcome, in cancer patients will possibly result in the development of new diagnostic protocols in specific types of cancer.

The generic PI3K have greatly facilitated the deciphering of the biological responses and signaling pathways controlled by PI3K and has even proved to be successful in a variety of pre-clinical models of human cancer.

The new generations of pharmacological inhibitors with increased specificity for one or a restricted number of PI3K isoforms and related kinases are entering clinical trials. It has been demonstrated that combining these agents with receptor tyrosine kinase inhibitors, which have already been evaluated protects p53 protein in the survival signals, permitting the p53 gene to function as a guardian of the genome.

Such information is in our opinion of great importance, since there exist, in addition to the well-studied class I PI3K isoforms, three class II enzymes, with still poorly understood cellular functions, at least in human cancer. In addition, the class III PI3K Vps34p should also be taken into account when designing future therapies based on targeting PI3K isoforms, in view of its roles in autophagy and intracellular trafficking events.

1.1. Methods of detection of CLL with deletion of p53 gene by FISH technology

The cause of CLL is unknown. There is no link to radiation. It is unclear if certain chemicals can cause CLL. CLL usually affects older adults, especially those over age 60. People under age 45 rarely develop CLL. CLL is more common in whites than in other ethnic groups. It is more common in men than in women. Some people with CLL have family members with the disease. Tests to diagnose CLL include: complete blood count (CBC) with blood cell differential, bone marrow biopsy, CT scans of the chest, abdomen, and pelvis, immunoglobulin testing, and the flow cytometry test of the white blood cells.

The diagnosis of CLL requires the presence of 109/LB lymphocytes in the peripheral blood, sustained for at least 3 months. The clonality of these B lymphocytes needs to be confirmed by demonstrating immunoglobulin light chain restriction using flow cytometry.

The leukemia cells found in the blood smear are characteristically small, mature lymphocytes with a narrow border of cytoplasm and a dense nucleus lacking discernable nucleoli and partially aggregated chromatin. Gumprecht nuclear shadows, or smudge cells, found as cellular debris, are additional morphologic features commonly associated with CLL.

Finding 55%prolymphocytes would favor a diagnosis of prolymphocytic leukemia; however, this diagnosis remains difficult and is solely based on morphological criteria, because no reliable immunological or genetic marker has been identified. A significant proportion of circulating prolymphocytes (10%) seems to indicate a more aggressive form of CLL (with NOTCH1 or genetic p53 aberrations).

1.2. Molecular genetics interphase fluorescence in situ hybridization

Our experience was evaluated with a FISH CLL panel after it had been implemented by clinicians at NMH to assess prognosis. We found that the FISH results were also useful for clarifying the diagnosis in cases in which the phenotype and/or morphologic features were atypical. This was especially true for cases with chromosome 14q32 aberrations.

Two cases were classified as Manta cell lymphoma (MCL) after FISH analysis showed CCND1/IGH fusion, indicating t, (11;14), (q13; q32). Both patients had blood lymphocytosis and some immunophenotypic features that resembled CLL. Both cases were CD20+, CD5+, and CD23+ (one bright and one dim). However, both cases had atypical morphologic features for CLL and were also FMC7+ and CD79b+, features that are unusual for CLL. The FISH results confirmed the diagnosis of MCL.

The Bcl-11a gene codes for a transcription factor that is expressed in many hematopoietic tissues, including bone marrow cells and germinal center B cells, and is required for normal lymphoid development. The bcl-11a translocation is extremely rare in CLL. Prior reported cases include the initial description of 2 children with CLL. Both children died 1 to 4 years after receiving bone marrow transplants for progressive CLL. Two adults with CLL and bcl-11a translocations have also been described.

Both had an aggressive clinical course and died with progressive disease. Our patient with a bcl-11a translocation also had a relatively aggressive course characterized by massive generalized lymphadenopathy, splenomegaly, and extensive bone marrow involvement associated with severe thrombocytopenia and anemia. His disease showed no response to chemotherapy, and he died 51 months after diagnosis.

Deletion of chromosome 11q was also present in the patient with a bcl-11a translocation. Chromosome 11q deletion is associated with extensive lymphadenopathy and a more advanced stage of CLL. The extent to which chromosome 11q deletion contributed to the aggressive clinical behavior is unclear. The prior reported CLL cases with bcl-11a translocations also had complex karyotypes that included chromosome 11q deletion in at least 1 adult.

Recognition of a Bcl-11a translocation occurring in CLL cases is important for at least 2 reasons. First, as in our case, the morphologic features of the lymphocytes may suggest non-Hodgkin's lymphoma, making it difficult to establish a definitive diagnosis of CLL based on analysis of only peripheral blood samples and bone marrow trephine biopsy sections. This is particularly true when proliferation centers are not present in the trephine biopsy section. Although the immunophenotype of the lymphocytes may be typical for CLL, a lymph node or other extramedullary tissue biopsy may be necessary to confirm the diagnosis. Second, because CLL cases with the Bcl-11a translocation seem to be clinically aggressive, their recognition may allow for initiation of early and possibly innovative therapy.

The bcl-2 translocation in our patient was conventional in that it involved the IGH gene. However, more detailed molecular analysis was not performed to determine the precise breakpoint within the bcl-2 gene. FISH can be performed with peripheral blood lymphocytes and identifies cytogenetic lesions in 80% of all CLL cases [15]. The most common deletions are in the long arm of chromosome 13 (del(13q)).

Additional, frequent chromosomal aberrations comprise trisomy of chromosome 12 and deletions in the long arm of chromosomes 11 (del(11q)) and in the short arm of chromosome 17 (del(17p). FISH represents a method for highlighting chromosomal anomalies, more sensitive than conventional karyotyping, allowing the analysis of chromosomes both in interphase and metaphase. The method is a genetic analysis technique (recombinant DNA technology) and consists in coupling a fluorescently labeled nucleic acid sample with a specific chromosomal region (classic in situ hybridization with digoxigenin or fluorescence).

Fluorescence in situ hybridization (FISH) is a molecular cytogenetic technique used to identify chromosomal, numerical and structural abnormalities (which cannot be detected by karyotyping). The principle of this method consists in attaching to the target sequence a single-stranded DNA probe (about 40 kb), fluorescently labeled, based on complementarity, with a target sequence of a chromosome.

The hybridization of the probe with the cellular DNA is visualized under a fluorescence microscope equipped with excitation and emission filters, which allows the reading of signals specific to the target area. Signals present in one hundred cells are counted and analyzed (for example, the three specific fluorescent signals for chromosome 21 indicate Down syndrome).

The absence of the fluorescent signal corresponding to the probe on a certain chromosome means the deletion (absence) of the respective area. The appearance of two different signals corresponding to the probe indicates microduplication. The probes are also used to identify chromosomal breakpoints in the case of translocations (structural anomalies that involve an exchange of fragments between two or more chromosomes) **[Photo 6]**.

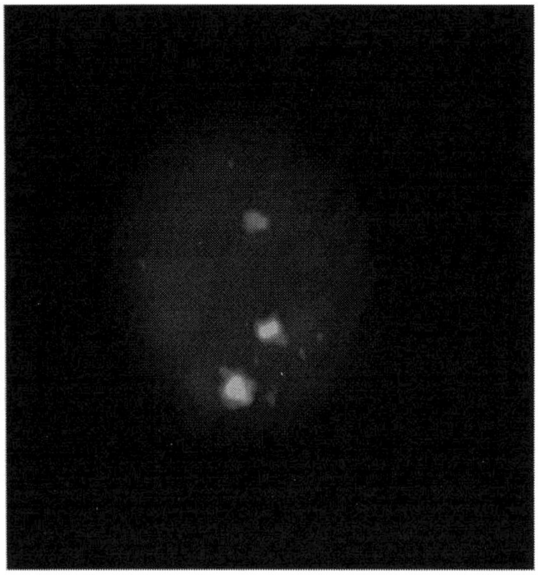

Photo 6. *Absence of a specific signal marked with red fluorochrome (TP53 locus) and the presence of two specific signals marked with green fluorochrome (ATM locus), model compatible with the deletion of the TP53 locus.*

Such probes can be useful in the diagnosis of syndromes produced by chromosomal microdeletions or microduplications: Williams, Beckwith-Wiedemann, Prader-Willi, Angelman, velo-cardio-facial (Di George, 2001), cat-eye, etc. The phenotype of these syndromes includes facial dysmorphia, mental retardation and various malformations.

The centromeric probes have a role in the identification of a certain chromosome and/or the detection of numerical anomalies of the respective chromosome; Telomere probes allow the diagnosis of microdeletions or other cryptic rearrangements; Sub-telomeric probe. Probes specific to a chromosome allow the "coloring" of the entire chromosome or only a region. Through the different coloring of two or more chromosomes, small rearrangements in which they are involved can be identified **[Scheme 22]**.

Gene Amplification / Gene Loss

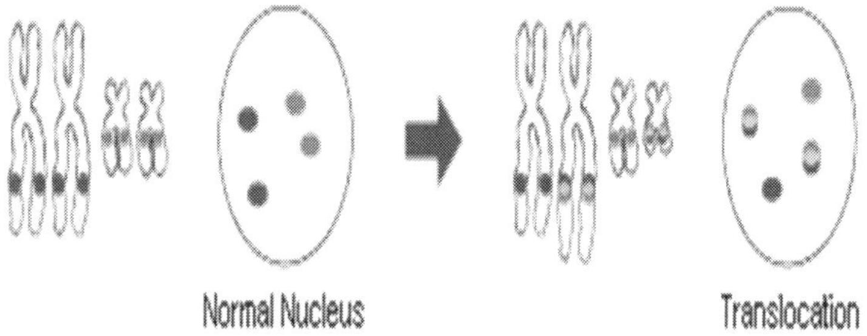

Split Dual Color FISH Probe

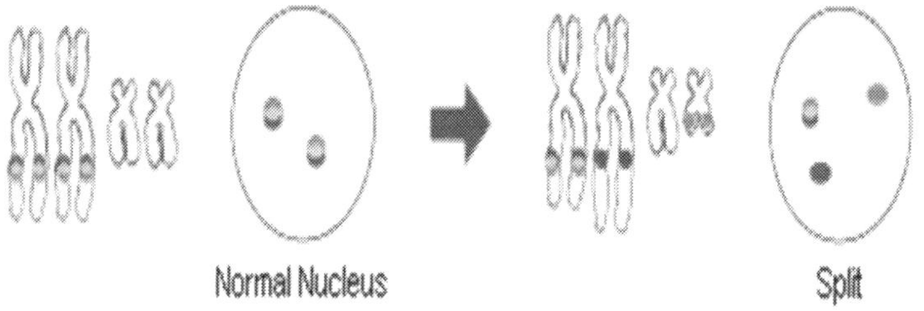

Scheme 22. *The probe for p53 (17p13.1) allows the identification of deletion of the p53 gene. Approximately 17% of B-cell CLL cases will have this deletion. The presence of p53 deletion is a poor prognostic marker for patients with CLL.*

Biological targets related to CLL starting from Bruton tyrosine kinase

BTK was placed in the signal transduction pathway downstream of the B cell receptor (BCR). This receptor is expressed on the B cell surface and has the unique capacity to specifically recognize antigens due to hypervariable regions present in the immunoglobulin heavy (IGH) and light (IGL) chains that together form the BCR.

Since myeloid cells are important components of the tumor microenvironment and particularly tumor-associated macrophages contribute to cancer progression, there is currently considerable interest in BTK inhibition as an anti-cancer therapy not only in B cell leukemias but also in other hematological malignancies and solid tumors.

BTK is essentially cytoplasmic and is only transiently recruited to the membrane through the interaction of its PH domain with phosphatidylinositol-3,4,5-triphosphate (PIP3), which is generated by phosphatidylinositol-3 kinase (PI3K).

BTK activation occurs in two steps upon its recruitment to the cell membrane. BTK-deficient B cells enter early G1, but not the S phase of the cell cycle, because they fail to induce cyclin D2 expression [30, 31, 32].

ATP-dependent Chromatin Remodeling Chromatin remodeling is the enzyme-assisted movement of nucleosomes on DNA. Chromatin is a condensed and often inaccessible structure where genomic DNA is packaged through histone and non-histone proteins. When DNA damage occurs, efficient and accurate repair of DNA damage ensures genome stability and prevents damage development which could lead to cancer or cell death. Activating DNA damage response (DDR) enables the cells to utilize post-translation histone modifications and ATP-dependent chromatin remodeling to modulate chromatin structure and increase the accessibility of the repair machinery to lesions embedded in chromatin, Chromatin remodeling utilizes the energy of ATP to disrupt nucleosome DNA contacts, move nucleosomes along DNA, and remove or exchange nucleosomes such that DNA repair can be accomplished.

Via ATP hydrolysis, the chromatin structure of several large multiprotein complexes (200 kDa – 2 MDa) can be enzymatically modulated. Several chromatin remodeling complexes are involved in the process: switch/sucrose non-fermentable (SWI/ SNF) family containing either the

Brahma (BRM) or Brahma-related gene 1 (BRG1). ATPase which slide and eject nucleosomes, imitation switch (ISWI) complexes containing SNF2H or SNF2L ATPase and mediate nucleosome sliding and histone displacement, inositol requiring 80 (INO80) chromatin remodeling factors containing INO80 ATPase or related SWR1-like factors such as the p400 ATPase which features long insertion in the middle of the conserved ATPase domain, and chromodomain helicase DNA binding protein (CHD) family members containing two tandemly arranged chromodomains (CDs) on the N-terminus of their ATPase which is involved in binding methylated histone tails as well as DNA and can slide and eject histones and have both activator and inhibitory roles in transcription regulation.

An ATPase which is capable of DNA translocation moves nucleosomes such that transcription factors can access DNA. Luijsterburg and van Attikum recently linked chromatin and the DNA damage response with cancer.

A Janus-activated kinase (JAK)–STAT pathway leading to impaired transcriptional activity. Janus Kinase (JAK) protein tyrosine kinases are recognized as an integral component of the cytokine receptor subunits, and enzyme activation, as the initiating step in a signaling cascade required for embryonic development, tissue growth, haemopoietic development and differentiation, innate and adaptive immunity and the inflammatory response. Given that JAK-STAT signaling is the universal and essential intracellular pathway for cytokine action, several regulatory mechanisms have evolved to control the magnitude and duration of signaling.

This allows fine-tuning of cytokine-mediated cellular effects and prevents the inappropriate activity often associated with disease development. There are three major mechanisms for negative regulation: receptor internalization, de-phosphorylation by phosphor-tyrosine phosphatases (PTPs), and direct inhibition by protein inhibitors of STATs (PIAS) and suppressor of cytokine signaling (SOCS) proteins.

While loss-of-function mutations in JAK3 and TYK2 are associated with immunodeficiency, the majority of naturally occurring mutations in JAK2, including chromosomal translocation, point mutations, insertions and deletions, are gain-of-function mutations and are associated with acute leukemia or myeloproliferative disorders (MPDs).

The chromosomal translocations of JAK2 loci (e.g. TEL-JAK2, PCM1-JAK2, BCR-JAK2 and PAX-JAK2) lead to the development of both myeloid and lymphoid hematological malignancies. These translocations result in the fusion of the JAK2 catalytic kinase (JH1) domain with multimerization subunits of partner proteins leading to constitutive tyrosine kinase activity and transformation.

Given the involvement of the V617F mutation in MPNs, most of the therapeutic development has focused on identifying low molecular mass ATP-competitive JAK2 inhibitors, reviewed in (296, 297). TG101348 and Ruxolitinib (or INCB018424) (298, 299) are two of many selective and potent JAK inhibitors currently in use in phase II and III MPN clinical trials, respectively. Ruxolitinib has a strong inhibitory effect on both JAK1 and JAK2, while TG101348 has greater selectivity for JAK2. Both of these inhibitors effectively attenuate downstream signaling, including phosphorylation of STAT and ERK1/2, and induce apoptosis in vitro. MPN patients treated with these inhibitors demonstrated a dramatic improvement in constitutional symptoms and a reduction in spleen size.

Although unpleasant side effects and a less satisfactory improvement in cytopenia, marrow fibrosis and JAKV617F burden were also observed, inhibition of JAK2 has already proven to be an excellent target for therapeutic intervention in MPNs.

AK-STAT signaling can interconnect with other cell-signaling pathways, such as the PI3K/AKT/mTOR pathway. When JAKs are activated and phosphorylate tyrosine residues on receptors, proteins with SH2 domains (such as STATs) can bind to the phosphor-tyrosine, and the proteins can carry out their function. Like STATs, the PI3K protein also has an SH2 domain, and therefore it is also able to bind to these phosphorylated receptors. As a result, activating the JAK-STAT pathway can also activate PI3K/AKT/mTOR signaling [34].

JAK-STAT signaling can also integrate with the MAPK/ERK pathway. Firstly, a protein important for MAPK/ERK signaling, called Grb2, has an SH2 domain, and therefore it can bind to receptors phosphorylated by JAKs (in a similar way to PI3K). Grb2 then functions to allow the MAPK/ERK pathway to progress. Secondly, a protein activated by the MAPK/ERK pathway, called MAPK (mitogen-activated protein kinase), can phosphorylate STATs, which can increase gene transcription by STATs. However, although MAPK can increase transcription induced by STATs, one study indicates that phosphorylation of STAT3 by MAPK can reduce STAT3 activity. One example of JAK-STAT signaling integrating with other pathways is Interleukin-2 (IL-2) receptor signaling in T cells.

IL-2 receptors have γ (gamma) chains, which are associated with JAK3, which then phosphorylates key tyrosine on the tail of the receptor [35].

Phosphorylation then recruits an adaptor protein called Shc, which activates the MAPK/ERK pathway, and this facilitates gene regulation by STAT5. Given that many JAKs are associated with cytokine receptors, the JAK-STAT signaling pathway plays a major role in cytokine receptor signaling. Since cytokines are substances produced by immune cells that can alter the activity of neighbor cells, the effects of JAK-STAT signaling are often more highly seen in cells of the immune system. For example, JAK3 activation in response to IL-2 is vital for lymphocyte development and function. Also, one study indicates that JAK1 is needed to carry out signaling for receptors of the cytokines IFNγ, IL-2, IL-4 and IL-10 [36].

The JAK-STAT pathway in cytokine receptor signaling can activate STATs, which can bind to DNA and allow the transcription of genes involved in immune cell division, survival, activation and recruitment. For example, STAT1 can enable the transcription of genes which inhibit cell division and stimulate inflammation. Also, STAT4 can activate NK cells (natural killer cells), and STAT5 can drive the formation of white blood cells [37].

In response to cytokines, such as IL-4, JAK-STAT signaling is also able to stimulate STAT6, which can promote B-cell proliferation, immune cell survival, and the production of an antibody called IgE. AK-STAT signaling plays an important role in animal development. The pathway can promote blood cell division, as well as differentiation (the process of a cell becoming more specialized). In some flies with faulty JAK genes, too much blood cell division can occur, potentially resulting in leukemia JAK-STAT signaling has also been associated with excessive white blood cell division in humans and mice.

Mutations in JAK2 can lead to leukemia and lymphoma. Specifically, mutations in exons 12, 13, 14 and 15 of the JAK2 gene are proposed to be a risk factor in developing lymphoma or leukemia. Additionally, mutated STAT3 and STAT5 can increase JAK-STAT signaling in NK and T cells, which promotes very high proliferation of these cells, and increases the likelihood of developing leukemia.

Also, a JAK-STAT signaling pathway mediated by erythropoietin (EPO), which usually allows the development of red blood cells, may be altered in patients with leukemia [38].

1.1. Phosphoinositide 3-kinase (PI3K)/AKT pathway resulting in increased apoptosis

The phosphoinositide 3-kinase (PI3K) pathway is frequently activated in human cancer and represents an attractive target for therapies based on small molecule inhibitors. PI3K isoforms play an essential role in the signal transduction events activated by cell surface receptors including receptor tyrosine kinases (RTKs) and G-protein-coupled receptors (GPCRs). There are eight known PI3K isoforms in humans, which have been subdivided into three classes (I-III). Therefore, PI3Ks show considerable diversity and it remains unclear which kinases in this family should be targeted in cancer.

The class IA of PI3K comprises the p110α, p110β and p110δ isoforms, which associate with activated RTKs. In human cancer, recent reports have described activating mutations in the PIK3CA gene encoding p110α, and inactivating mutations in the phosphatase and tensin homologue (PTEN) gene, a tumor suppressor and antagonist of the PI3K pathway. The PIK3CA mutations described in cancer constitutively activate p110α and, when expressed in cells drive oncogenic transformation.

Moreover, these mutations cause the constitutive activation of downstream signaling molecules such as Akt/protein kinase B (PKB), mammalian target of rapamycin (mTOR) and ribosomal protein S6 kinase (S6K) that is commonly observed in cancer cells. In addition to p110α, the other isoforms of the PI3K family may also play a role in human cancer, although their functions remain to be precisely identified. In this review, we will discuss the evidence implicating individual PI3K isoforms in human cancer and their potential as drug targets in this context. In addition, PI3K was reported to be involved in the activation of several other protein kinases, including c-Jun N-terminal kinase (JNK) by EGF [151]. Bruton's tyrosine kinase (Btk), which has a pH domain, can bind PIP3 [39].

RNA interference (RNAi) as well as conditional knock-out of ILK had no effect on phosphorylation of Akt on Thr-308 but resulted in almost complete inhibition of phosphorylation on Ser-473 and significant inhibition of Akt activity, accompanied by significant stimulation of apoptosis. In addition, Raf-1 activation by Ras was shown to be achieved through a combination of both physical interaction and indirect mechanisms involving the activation of PI3K as a second Ras effector, which directs p21-activated kinase (PAK)-mediated regulatory phosphorylation of Raf-1 [40].

Phosphorylation of Raf-1 on Ser338 through PI3K and Pak was also shown to provide a co-stimulatory signal which together with Ras leads to strong activation of Raf-1 kinase activity by integrins. All cycle progression is a tightly controlled process. To initiate cell division, mitogens trigger several early signals that promote the G(0)-G(1) transition by inducing cell growth and the activation of G(1) cyclins. Activation of cyclin E/cdk2 (cyclin-dependent kinase 2) at the end of G(1) is then required to trigger DNA synthesis (S phase entry). Among the early signals induced by mitogens, activation of PI3K appears essential to induce cell cycle entry, as it regulates cell growth signaling pathways (see previous section), which in turn determines the rate of cell cycle progression. A mechanism by which PI3K and its downstream effector Akt regulate cell cycle entry is by inactivation of the FOXO transcription factors, which induce expression of quiescence genes such as that encoding p27 (Kip), p130 and cyclin G2. PI3K/FOXO then work as a complementary switch: when PI3K is active, FOXO transcription factors are inactive [41].

PTEN expression also led to the inhibition of Akt/ PKB. It was also shown that the PTEN protein induces a G1 block when reconstituted in PTEN-null cells. A PTEN mutant associated with Cowden's disease (PTEN; G129E) has protein phosphatase activity yet is defective in dephosphorylating inositol 1,3,4,5-tetrakisphosphate in vitro and fails to arrest cells in G1. These data suggest a link between the induction of a cell-cycle block by PTEN and its ability to dephosphorylate, in vivo, PIP3. In a parallel study, PTEN impaired the activation of endogenous Akt in cells and inhibited the phosphorylation of 4E-BP1. In addition, PTEN/ MMAC1 repressed gene expression in a manner that is rescued by Akt but not PI3K [42].

The PTEN tumor suppressor protein inhibits PI3K/Akt signaling that promotes translocation of Mdm2 into the nucleus. When restricted to the cytoplasm, Mdm2 is degraded. The ability of PTEN to inhibit the nuclear entry of Mdm2 increases the cellular content and transactivation of the p53 tumor suppressor protein. Retroviral transduction of PTEN into PTEN null glioblastoma cells increases p53 activity and expression of p53 target genes and induces cell cycle arrest U87MG/PTEN glioblastoma cells were more sensitive than U87MG/PTEN null cells to death induced by etoposide, a chemotherapeutic agent that induces DNA damage.

These results established a direct connection between the activities of two major tumor suppressors and show that they act together to respond to stresses and malignancies. PTEN almost twenty years after the discovery of PI3K, this family of enzymes has become the focus of intensive research aimed at developing new drugs for a wide variety of diseases, including cancer, inflammation and allergy.

The discovery of the many genetic alterations affecting the PI3K/PTEN/Akt/mTOR pathway in human cancer has placed some of the enzymes involved, including p110α, mTOR and Akt, on the top list of druggable targets in the field of oncology research. Moreover, the recognition that some of the molecular alterations, such as increased Akt phosphorylation or PTEN mutations correlate with clinical parameters, such as outcome, in cancer patients will possibly result in the development of new diagnostic protocols in specific types of cancer.

The generic PI3K have greatly facilitated the deciphering of the biological responses and signaling pathways controlled by PI3K and has even proved to be successful in a variety of pre-clinical models of human cancer. The new generations of pharmacological inhibitors with increased specificity for one or a restricted number of PI3K isoforms and related kinases are entering clinical trials. It has been demonstrated that combining these agents with receptor tyrosine kinase inhibitors, which have already been evaluated protects p53 from survival signals, permitting p53 to function as a guardian of the genome [43, 44].

Such information is in our opinion of great importance, since there exist, in addition to the well-studied class I PI3K isoforms, three class II enzymes, with still poorly understood cellular functions, at least in human cancer. In addition, the class III PI3K Vps34p should also be taken into account when designing future therapies based on targeting PI3K isoforms, given its roles in autophagy and intracellular trafficking events. Telomere probes allow the diagnosis of microdeletions or other cryptic rearrangements; Sub-telomeric probe. Probes specific to a chromosome allow the "coloring" of the entire chromosome or only a region. Through the different coloring of two or more chromosomes, small rearrangements in which they are involved can be identified **[Scheme 23]**.

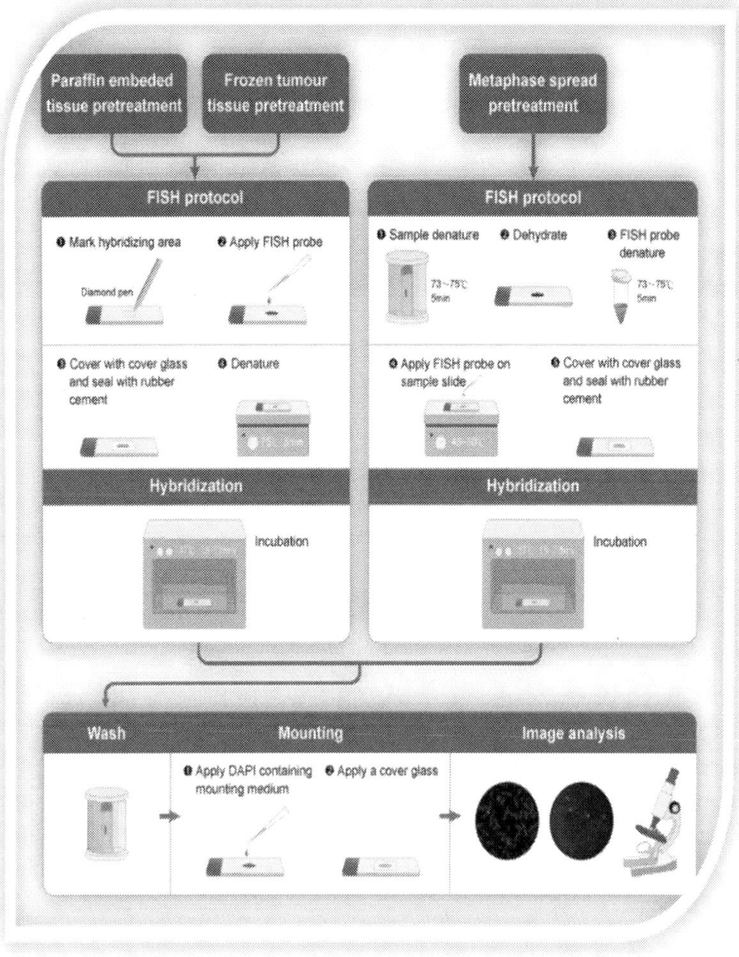

Scheme 23. *Work protocol of fixed FISH technique.*

Deletions of chromosomal region 13q14 were frequently associated with cases of CLL and lymphocytosis with monoclonal B cells may precede occasionally CLL or aggressive lymphomas and is suggested that this region contains a tumor suppressor gene, similar to gene p53 which combines their work. All this leads us to conclude that the appearance of

CLL deficiency occurs through a common p53 gene and chromosomal region 13q14 gene.

CLL and Hodgkin's lymphoma (HL) are particularly dependent on their microenvironment and have associated signaling pathways and deletion of miR15/16 locus, common especially, in CLL. Micro-ARN, miR15 and miR16 are located at chromosome 13q14, a region deleted in more than half of B-cell chronic lymphocytic leukemia (B-CLL) cases. The micro-RNA deletions, miR-15a/16-1, both accelerate the proliferation of human B cells by modulating the expression of genes that control cell cycle progression. These results define the role of the 13q14, with deletions of p53 in the pathogenesis of CLL.

Prospective clinical trials indicate that certain genetic abnormalities are associated with adverse outcomes in response to standard chemo-immune therapy regimens. Patients with leukemia cells that carry del(17p) and/or p53 gene mutations (as determined by DNA sequencing, with a cutoff of 10%) have an inferior prognosis and appear relatively resistant to standard chemotherapy regimens using alkylating drugs and/or purine analogues.

Because additional genetic abnormalities may be acquired during the disease, genetic analyses (in particular for del(17p)/TP53 mutations) should be repeated before any subsequent second or third-line treatment. It has been demonstrated that the progression-free survival and overall survival of CLL patients carrying a del(17p) and patients carrying a p53 mutation, as detected by Sanger sequencing in the absence of del(17p). If the specific patient subgroup is not excluded, then they should be stratified or analyzed as a subgroup analysis based upon whether they have leukemia cells with del(17p) or del(11q), and for mutations of the p53 gene [45, 46].

The frequencies of p53 gene mutations, deletions or translocations, in CLL, can be categorized as the individual biomarkers in proteomic and genomic profiles for this type of leukemia and can be implemented in choices of targeted treatments from personalized medicine. The most common form of therapy using DNA encoding is using a functional gene to replace a mutated gene.

The polymer molecule is packaged in a "vector" molecule within the cell bearing. Bidirectional Sanger sequencing of all protein-encoding exons of TP53 was performed using peripheral blood lymphocyte DNA from breast/ovarian cancer probands from 37 HBC/HBOC Canadian families.

Five new p53 mutations were identified in six pedigrees from hereditary cancer clinics. No deleterious mutations were identified in cancer probands from 37 HBC/HBOC families. A targeted mutation screen of the 1235 breast cancer cases identified a c.844C>T [p. Arg282Trp] mutation carrier. This mutation was also found among the six mutation-positive cancer families provided by the local hereditary cancer clinics. The targeted screen also uncovered a new p53 mutation, c.685T>C [p. Cys229Arg], in two breast cancer cases. All TP53 mutation carriers were among the 656 women with breast cancer diagnosed at less than 50 years of age (BMC Med Genet 2015).

It provided a possible explanation for the inability of BCR signals to induce the entry of CLL cells into the S phase of the cell cycle and, second, it raised the possibility that certain genetic lesions associated with CLL progression may function by reducing the requirement for costimulatory signals in driving CLL proliferation. Such lesions most notably include deletions of the CDKN2A/2B locus and mutations/ deletions of the CDKN1A regulator TP53, which co-occur in approximately one-quarter of cases with clonally related Richter syndrome (RS), an aggressive B-cell lymphoma arising from the transformation of CLL B cells. Recent research included this method to treat CLL.

Recent studies have focused on the mechanisms of leukemic progression associated with mutant p53 gene, a tumor suppressor gene that is commonly mutated and is associated with adverse outcomes. In myeloid neoplasms, the presence of a p53 gene mutation is associated with complex cytogenetic abnormalities and dire outcomes. In the leukemic phase of myeloid leukemia, p53 gene mutations are present in one-third of cases but small p53 clones are found in a sizable cohort (15%) of chronic phase MPN86 that do not transform to acute myeloid leukemia (AML).

The biallelic loss of wild-type p53 is frequent in leukemic phase MPN87 and other advanced myeloid neoplasms, suggesting that multiple mechanisms are required for leukemia development. In a recent study, ~15 000 single HSPCs were analyzed in 14 advanced phase MPN cases. In all cases, a biallelic loss of the wild-type P53 gene was observed in the principal clone. The "multi-hit" mutant HSPCs contained 2 broad cellular subsets; one showing an erythroid gene signature, and the other enriched in leukemia stem cell genes.

The finding of expanded erythroid-biased HSPCs was not expected, as leukemias secondary to MPNs are not typically erythroleukemia according to standard clinical criteria [26]. Wild-type patient HSCs showed a reduction in cell cycle genes, suggestive of enhanced quiescence and increased expression of inflammatory genes, leading to the hypothesis that a cell-extrinsic inflammatory process mediates disease evolution and exerts a fitness advantage to the p53 gene mutant clone while repressing wild-type HSPCs, potentially through DNA damage-related mechanisms. This study is another example of how single-cell studies applied to MPN can have a broad relevance for human diseases, given that p53 is the most commonly mutated gene in human cancers.

CHAPTER NINE

REACTIVATION OF ISOFORM P53 PROTEIN BY PEPTIDES

Peptides have been screened to identify candidates which can favor the correct conformation of p53 protein, thereby restoring p53 protein functions. These peptides, when transfected in cell lines, displayed enhanced cell death and the DNA binding ability of p53 protein. Tumor xenograft assays showed complete suppression of the tumor in some cases and reduction in tumor size in others.

Another synthetic peptide (peptide 46), corresponding to the C-terminal residues of p53 (361–382), has shown reactivation of and restoration of p53 protein transcriptional activity in some mutant p53 genes such as R273H. When transfected in Saos-2 cells, the peptide expressing an isoform p53 R273H protein displayed cell-cycle inhibition and apoptosis.

Furthermore, similar effects on other human cancer cell lines expressing mutant or wild-type p53 proteins were also observed in the presence of peptides. A nine-residue peptide, CDB3, was shown to bind the core domain of p53 protein, leading to its stabilization under in vitro conditions. CDB3 can restore several p53 mutants to wild-type conformation, ensuring they retain their DNA binding ability.

Previous reports in breast cancer and ovarian cancer cell lines have suggested that PRIMA-1 can reverse isoform p53 protein aggregate accumulation in cancer cells by substantially decreasing p53 aggregates.

Isoform p53-expressing Saos-2 cells also show selective growth-inhibitory and apoptosis-inducing effects when treated with PRIMA-1 (2,2-bis(hydroxymethyl)-1-azabicyclooctan-3-one). Furthermore, PRIMA-1 and its structural analogue PRIMA-1MET (APR-246) can also inhibit human xenograft tumor growth in SCID mice and PRIMA-1 can therefore restore the sequence-specific DNA binding and can activate the functional conformation of mutant p53 in vitro as well as in living cells. PRIMA-1 was observed to rescue both contact and structural p53 mutants.

Like PRIMA-1, the maleimide-derived molecule MIRA-1 can also reactivate the DNA binding property of isoform p53 protein in vitro as well as in cells. In SCID mice, a structural homolog of MIRA-3 had anticancer efficacy against human tumor xenografts expressing isoform p53 protein.

CRISPR-Cas9 systems provide a platform for direct and highly-efficient modification of genomic DNA. However, the delivery of Cas9 and the synthesis of the guide RNA (gRNA) remain limiting factors in the gene editing workflow. The introduction of gRNA ribonucleoprotein (RNP) complexes has enabled the effective delivery of CRISPR editing tools for various mammalian cell engineering applications, whilst overcoming the existing challenges with traditional approaches such as plasmids.

While not all Cas9 enzymes can be utilized in the RNP format, Cas9 proteins are gaining popularity due to their promising predictability, efficiency, and robustness in delivery and editing, along with overall ease of use and time savings.

Sequences do not need to be cloned into a plasmid, and the mixing of the Cas9 protein and gRNA is quick and easy. It is also possible to order multiple synthetic gRNA sequences that target various genes simultaneously.

The use of Cas9 proteins limits the chances of off-target edits and toxicity. Plasmids cause constant expression of Cas9, which can lead to many off-target edits along with the integration of the plasmid into random locations in the genome, which can in turn result in further mutations and changes. Cas9 proteins pose no such risk – even at increased concentrations – which is of particular importance for researchers working with cells such as embryonic stem cells that are intolerant to plasmid transfection.

Cas9 proteins can be directly introduced into cells, streamlining the workflow and saving time without compromising efficiency and efficacy. Once Cas9 proteins are complexed with gRNA, they are easily delivered into cells.

This method has shown superior genome editing for more challenging cell lines such as induced pluripotent stem cells (iPSCs) or primary T-Cells as well as offering the editing of multiple targets. Being DNA-free, transcription or translation processes are not required, meaning the protein is available for use immediately. Due to this quick turnover, there is a lower chance of off-target effects occurring.

To address the possibility that genetic lesions in CDKN2A/2B and TP53 cooperate with BCR signals in the pathogenesis of RS, we used clustered and regularly interspersed short palindromic repeats of (CRISPR)/ CRISPR-associated protein 9 (Cas9) technology to simultaneously target the CDKN2A, CDKN2B, and TP53 genes in primarily Eμ-TCL1-derived murine CLL cells expressing different autoreactive BCRs [11]. We show that the genetically modified leukemia cells acquire the capacity to proliferate spontaneously in vitro but remain dependent on BCR signals **[Scheme 24]**.

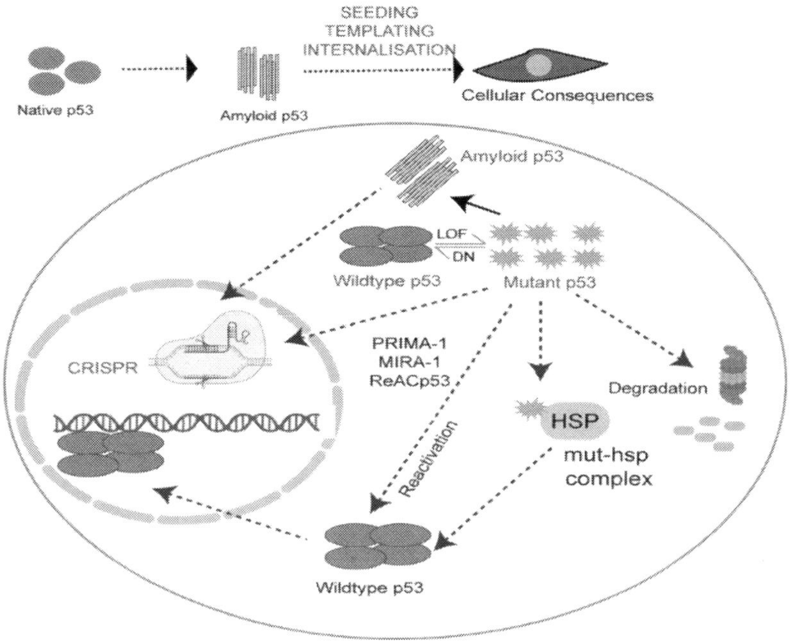

Scheme 24. *CRISPR-Cas9 systems provide a platform for direct and high-efficiency modification of genomic DNA.*

For example, a patient with a history of Chronic Lymphocytic Leukemia, treated with chemotherapy, was discovered a Diffuse Non-Hodgkin's

Lymphoma, with Richter Syndrome. The destruction of the normal lymph node architecture and the invasion of the capsule and the adjacent fatty tissue by the characteristic neoplastic cells constituted the histological criteria for PET / CT imaging analysis.

Bone marrow biopsy (M.O) to determine spinal cord locations should be performed in all patients. The primary extra-ganglionic involvement was the subject of clinical studies, so there was an increase in the incidence of extra-nodal non-Hodgkin's lymphomas in 20-30% of cases, the most common locations being in soft tissues. Approximately 5–10% of CLL patients receiving ibrutinib develop Richter syndrome, usually within 2 years of treatment, suggesting that this drug selects a pre-existing transformed clone **[Figure 2]**.

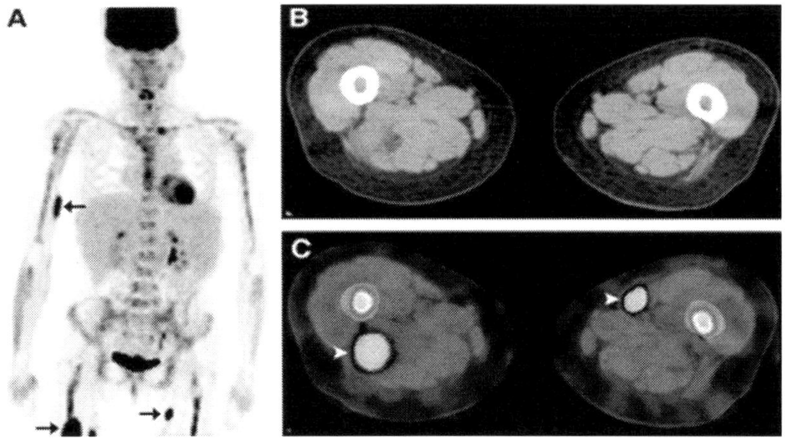

Figure 2*. PET image with maximum coronal intensity projection (A) demonstrated focal areas of increased absorption of 18F-FDG in the right forearm and both upper thighs (arrows). (B); hypermetabolic intramuscular masses (arrowheads) which were evident on fused axial PET / CT images (C); fine needle aspiration of a right forearm nodule revealed diffuse large cell lymphoma.*

p53 genes have been shown to promote hematopoietic stem cell stagnation (HSCs) and self-renewal. And recent studies have shown that p53 protein deficiency promotes CLL in cases of Richter syndrome, non-Hodgkin's lymphoma and Diffuse Large Lymphoma (DLL), by eliminating its ability to limit aberrant self-renewal in hematopoietic progenitors.

Small peptides such as the PRIMA1 peptide and CDB3 are effective in restoring p53 function. It has been suggested that PRIMA1 can restore the binding activity of the p53 protein to the DNA chain. Research has also shown that acetylation, deacetylation and phosphorylation of non-mutant p53 protein can repair DNA damage in cancer cells, but different isoforms of p53 protein have different cellular mechanisms in cancer.

The altered activity of the p53 protein in the isoform state affects DNA damage and even accelerates forms of cancer from the mild form to the severe form. Replacing the amino acid serine in the p53 protein with the amino acid alanine resulted in preventing the p53 protein from inhibiting the cell cycle. Research has shown that restoring the function of the isomorphic p53 protein could lead to the regression of certain cancer cells without damaging other cells in the process.

Nutlin-3, a cis-imidazole analogue, was found to strongly induce apoptosis in cell lines derived from hematologic malignancies and B-cell translocation LLC with frequent translocation of 14q32-17p, with a good therapeutic response. Other studies have suggested that a high concentration of adenosine triphosphate (ATP) in malignant B-cell lymphocytes in CLL affected the p53 gene to induce cell apoptosis.

It is recommended that before initiating treatment in CLL, ATP levels related to the presence or absence of p53 gene mutations should be measured to adjust doses of cytostatic treatments that have been resistant, causing the disease to relapse. To avoid very high doses of cytostatic, which can induce side effects, the administration of drugs such as AZD2014 (Vistusertib) and ATP-competitive or ATP inhibitory analogues, e.g. isopentenyl ester of ATP (Appl), which is a triphosphoric acid ester and inhibits the signals of the uncontrolled cell division of the cell cycle, are recommended in relapses from CLL. The respective researches are in stage I clinical trials.

Also, the treatment of CLL with the drug AZD8055 showed a significant reduction of cell proliferation and an increase of apoptosis. WYE-354, a heterocyclic compound, is a double-competitive ATP kinase inhibitor that selectively blocks TORC1 / 2 m activity.

In vitro, analyses have shown that WYE-354 strongly blocks the proliferation of cancer cells. PKI-587 drug, another competitive double inhibitor of ATP, is a therapeutic target for PI3K (PI3K-α and PI3K-γ) and mTOR, is an orally bioavailable inhibitor and also has inhibitory activity against the most mutant conditions of most frequent oncogenes.

Consistent with this role in the autophagy process, p53 activity is compromised in a large proportion of all cancers, either by mutation of the p53 gene (p53 coding) or by alterations of the statuses of p53 modulators. The contextual role of autophagy in cancer, which could be changed by p53 status, is expected to be developed in a new anti-cancer therapeutic approach.

In recent years, much attention has been paid to the importance of the p53-expressed protein in CLL. Non-Hodgkin's lymphoma (NHL) develops as a second malignancy in approximately 3-10% of patients with CLL. The increased expression of p53 protein concentrations in CLL-B is associated with poor responses to conventional chemotherapy.

In the context of a heterogeneous condition such as CLL-B, the ELISA method, which is cheap and safe, seems to provide a useful prognostic tool capable of identifying patients who can be considered candidates for therapeutic, targeted and personalized strategies **[Table 4]**.

Lymphocyte Remak µ Moli ATP in 10^6 cell/m	**Mean values (x)**	**The variation (s2)**	**Standard deviation (s)**	**Standard Error (s1)**	**Medin M**	**Derivation Coefficient (Cv)**
Normality	1.7476	0.1701535	0.4124966	0.0842003	1.72	23.6036%
Allergy	3.9160	0.3226356	0.5680102	0.1159443	4.09	25.5363%
Malignancy	3.8390	0.2152685	0.4639703	0.1546567	3.87	12.0857%

Table 4*. Apoptosis is an energy-dependent process, in contrast to necrosis, which also occurs in the absence of ATP, [Udristioiu A, Cristina F, Andrei PM. "High Concentration of anaerobic ATP implicated in aborted apoptosis from CLL". Lab Medicine, American Journal of Clinical Pathology-ASCP, Manuscript 09-08-LM-S-SCI-0122R1. Published in 04-05/2010, Indexed in Web of Science].*

Total results, in mean values of µ Moli ATP in 106 cell/ml measured method on Analyzer with luminescence: Clinically normal = 1.37µM; Bronchial asthma = 4.60 µM; Chronic urticaria = 3.70 µM Allergic rhinitis

= 2.77 μM; Lung neoplasm = 3.20 μM; Bone metastases = 3.67 μM; Chronic lymphatic leukemia = 4.42 μM.

Apoptosis is regulated by a cascade of proteins called caspases. Caspases are the apoptosis executor proteins and are present as pro-forms in all cells. After cleavage, caspases become active and initiate pathways leading to apoptosis. The pro-apoptotic proteins Bax, Bad, Bak and Bid induce programmed cell death. Proteins that can shift the balance towards survival are the anti-apoptotic proteins Bcl-2 and Bcl-Xl. An important regulator of apoptosis after DNA damage is the p53 protein. After DNA damage can induce the Bax and Bak proteins, which regulate the release of Cytochrome C, from the mitochondria, thereby initiating the cascade leading to apoptosis.

After the induction of apoptosis by hypoxia cytochrome C is released into the cytoplasm. Cytochrome C binds to the apoptotic protease activating factor 1 (Apaf-1). Apaf-1 activates caspase 9 which in turn cleaves caspases 3, then 6, 8, and 9, leading to cell death. The apoptosis can be induced in response to hypoxia. The severity of hypoxia determines whether cells become apoptotic or adapt to hypoxia.

An individual cell requires multiple signals to survive, otherwise, it will undergo programmed cell death. Respective of the initial signal, the core role belongs to a family of proteases that have cysteine at their active site and cleave target proteins at a specific aspartic site, called caspases. Since activation of caspases and their cleavage are irreversible, the entrance into apoptosis is in an all or non-mode, without a possibility to turn back.

Caspases are synthesized in inactive form as pro-caspases. Each caspase consists of structurally related molecules with a predominant, a large subunit and a small subunit. The two combine to form tetramers after cleavage. When activated, they cleave the aspartic moiety of another-caspases, mapmaking active. Currently, 13 different caspases have been described, and 11 of them were found in humans.

The leader position among upstream enzymes belongs to initiator procaspases-9. When they are pushed together by adapter proteins, they cleave themselves to active initiator caspase-9. In the enzymatic cascade, caspases activity is amplified until key reactions are reached: cleavage of protein that normally holds a DNase in an inactive form, cleavage of poly-ADP-ribose-polymerase, cleavage nuclear lamina. The most important regulatory mechanisms in mammalians are: death receptors, caspases, mitochondria, the Bcl-2 family proto-oncogenes and the tumor suppressor p53 gene.

The main death receptors (DRs) are CD95(Fas.), CD 120a, (TNF-R1), DR3, DR4, DR,5, and DR6, which are responsive to cytokines belonging to the tumor necrosis factor (THF-alpha, lymphotoxin, Fas-L and Apo-L3).

The link between apoptosis and TNF activity shows why abnormal production of TNF plays an important role in several autoimmune diseases: rheumatoid arthritis, multiple sclerosis, and diabetes mellitus.

Mitochondria have a fundamental position in executing apoptosis induced by intracellular signals. When cells are stressed due to physical signals, chemical stimulus, hypoxia or cytokines, mitochondria release pro-apoptotic proteins, including cytochrome C. It binds to an adapter protein Apaf-1, thus activating pro-caspase-9 and enzymatic cascade.

Additional mitochondria activities leading to apoptosis are the disruption of transport and modification of cellular reduction-oxidation potential. The most important proto-oncogenes involved in maintaining cell survival and proliferation are those belonging to the Bcl family. It is genes products have different functions: Bcl and Bcl-XL, are inhibitors of apoptosis, while Bad, Bax, Bak and Bid are promoters.

The protein Bad binds and inactivates inhibitors of apoptosis like Bcl-2 and Bcl-XL. The protein Bid is an activator of Bax and Bak, which like p53 protein are inhibitors of apoptosis through stimulating cytochrome C release from mitochondria. Among tumor suppressor genes, p53 is the most important. In response to DNA damage, the product of the p53 gene, protein p53, can induce apoptosis through activation of gene encoding cytochrome C proteins.

The p53 tumor-suppressor gene is mutated in 60% of human tumors, and the product of the gene acts as a suppressor of cell division. It is thought that the growth-suppressive effects of p53 are mediated through the transcriptional trans-activation activity of the protein. Overexpression of the p53 protein results either in arrest in the G (1) phase of the cell cycle or in the induction of apoptosis.

Both the level of the protein and its transcriptional trans-activation activity increase following treatment of cells with agents that damage DNA, and it is thought that p53 acts to protect cells against the accumulation of mutations and subsequent conversion to a cancerous state.

The induction of p53 levels in cells exposed to gamma-irradiation results in cell cycle arrest in some cells (fibroblasts) and apoptosis in others (thymocytes). Cells lacking p53 have lost this cell cycle control and presumably accumulate damage-induced mutations that result in tumorigenesis.

Thus, the role of p53 in suppressing tumorigenesis may be to rescue the cell or organism from the mutagenic effects of DNA damage. Loss of p53 function accelerates the process of tumorigenesis and alters the response of cells to agents that damage DNA, indicating that successful strategies for radiation therapy may well need to take into account the tissue of origin and the status of p53 in the tumor **[Scheme 25]**.

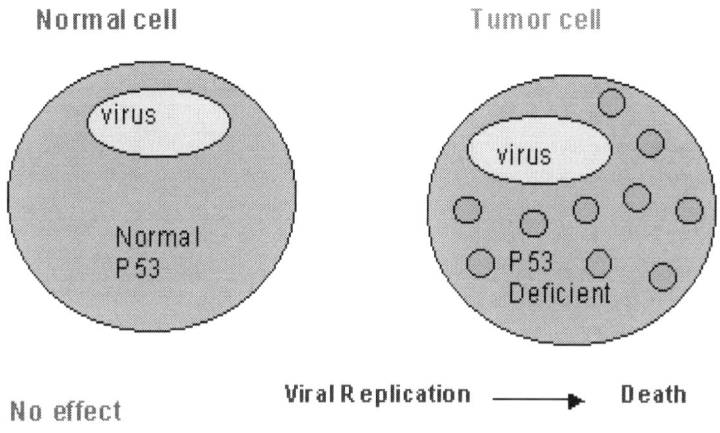

Scheme 25. *Loss of p53 function accelerates the process of tumorigenesis and alters the response of cells to agents that damage DNA.*

Hypoxia reduces the ability of cells to maintain their energy level because less ATP will be obtained from glycolysis than from oxidative phosphorylation. Cells will adapt by activating the expression of genes in glycolysis. In addition, cell proliferation and angiogenesis will be stimulated, enabling oxygenation of the tissues. Severe hypoxia causes a high mutation rate, resulting in point mutation, which may be explained by reduced DNA mismatch repair activity resulting from decreased MLH1 and PMS2, concentration which are caused by hypoxia.

In addition, hypoxia induces genetic instability by the induction of fragile sites causing gene amplification. During severe hypoxia or anoxia, the cells initiate a cascade of events that leads to apoptotic cell death with hypoxia-induced mutation. The supply of oxygen and nutrients is severely hampered by the malformed vessels. The combination of lack of oxygen and lack of nutrients causes energy deprivation.

Low ATP concentration in hypoxic tumor cells, disable the apoptotic cascade and induce cell death by necrosis. Severe hypoxia in the presence of energy stimulates cells to undergo apoptosis whereas oxygen levels above 0.5% prevent cell death. The most direct induction of hypoxia-induced apoptosis is the inhibition of the electron transport chain at the inner membrane of the mitochondria.

The lack of oxygen inhibits the transport of protons and thereby causes a decrease in the membrane potential. The reduction of mitochondrial-derived ATP causes activation of Bax or Bak, leading to cytochrome C release into the cytosol. Hence, fibroblasts of mice lacking Bax or Bak genes are resistant to oxygen deprivation.

Death resistance of hypoxia takes place on at least two levels in mitochondria cytosol.

Accumulation of Bax in the mitochondria caused the release of cytochrome C into the cytosol, which was strongly reduced in the hypoxic environment. This prohibited the cascade leading to cell death. Bax translocation was suppressed as a result of increased concentrations of the inhibitor of apoptosis protein 2, (IAP-2). In addition to the intrinsic apoptotic pathway, extrinsic pathways have been identified that can initiate and execute the cell death process.

Apoptosis via an extrinsic pathway is initiated by death ligands, such as the Fas ligand or the tumor necrosis factor (TNF), leading finally to the activation of caspases 8 or caspase 3, and resulting in activation of Apaf (activating factor protein). A third mechanism by which hypoxia can induce apoptosis is the activation of c-Jun NH2-terminal kinase (JNK), also termed stress-activated protein kinase (SAPC).

1.1. Wild-type JNK/SAPK had a slight pro-apoptosis effect

Both severe hypoxia and the presence of ATP are required to induce apoptosis. Oxygen levels above 0.5% prevent cells from undergoing apoptosis. Cell survival under mild hypoxia is mediated by phosphoinositide-3 kinase (PIK3) and its downstream target Akt.50.

The PI3K/Act.50 signaling pathway is important for cell survival and proliferation because it prevents Bad from inhibiting the anti-apoptotic activity Bcl-Xl. Act-induced NF-B (nuclear factor B), which leads to cell survival.

The key regulator of this process, hypoxia-inducible factor 1 (HIF-1) can initiate apoptosis by inducing high concentrations of pro-apoptotic proteins such as BNIp3 and can cause stabilization of p53 protein. During hypoxia anti-apoptotic proteins, such as IAP-2 (inhibitor of apoptosis protein 2), can be induced whereas the pro-apoptotic protein Bax can be down-regulated. Understanding the regulation of apoptosis during hypoxia and the mechanism of resistance to apoptosis might lead to more specific treatments for solid tumors. Hif-1 is a heterodimer composed of the rate-limiting factor HIF1 and the constitutively expressed HIF-1-beta. HIF-1-beta is also called the aryl-hydrocarbon receptor nuclear translocator. HIF1 is induced by hypoxia and also be oncogenes such as HER-2/neu, v-src, and ras.

The induction of HIF-1 by hypoxia takes place at the protein level because HIF-1 m RNA expression remains unchanged. During normoxia, HIF-1 is expressed but is unstable. Rapid degradation of HIF-1 by proteasome results from ubiquitination by the product of the Vo Hippel Lindau tumor suppressor gene (VHL). In patients where there is a loss of the VHL gene, HIF-1 and HIF-1-dependent genes, such as the angiogenesis factor, are also expressed during normoxia.

1.2. HIF-1 accumulates by hypoxia

Hypoxia-inducible factor 1 (HIF-1) accumulates when hypoxia can't be hydroxylated and is translocated to the nucleus. Here it binds to HIF-1-beta to form the active transcription factor HIF-1. HIF becomes unstable when bound to p53.Jabl-1(Jun activation domain-binding protein 1), directly interferes with the HIF-p53 complex and leads to the stabilization of HIF-1 protein during hypoxia.

Hypoxia in combination with poor diet and hypo-energy (hypoglycemia) reduced proliferation and increased apoptosis in wild-type embryonic cells but not in embryonic cell stem with inactivated HIF-1 genes. HIF-1 can induce apoptosis via two mechanisms: First, it can increase the stability of the product of the tumor suppressor p53 gene (protein p53).Then, in cases hwre environmental stress of the DNA damages p53, it induces programmed cell death by regulating proteins such as Bax, or it can cause growth arrest, which is mediated by p21. Recently, it was shown that HIF-1 directly binds to the p53 ubiquitin ligase MDM2, both in vivo and in vitro, thereby stabilizing the p53 protein.

However, another report showed direct binding of p53 to the COD domain of HIF-1, where HIF-1 interacts with wild-type p53 but not with tumor-derived mutant p53. HIF-1 can initiate hypoxia-mediated apoptosis by increasing the expression of Bcl-2 binding protein, (BNIP-3 and NIX), thereby inhibiting proteins the anti-apoptotic effect of Bcl-2, or by stabilizing wild-type p53; if the cell already has a p53 gene mutation, hypoxia-induced apoptosis is prevented.

Hypoxia by itself can also prevent apoptosis by inducing the expression of the ani-apoptotic protein IAP-2. Also, HIF-1 can induce apoptosis by the first mechanism and by twice mechanism can induce the expression of gene BNIP3 which may prove to be an important protein for the elimination of damaged cells by undergoing rapid cell death.

HIF-1 may also have an anti-apoptotic function because cells with high amounts of HIF-1 are more resistant to hypoxia-induced apoptosis. Because hypoxia-inducible factor 1(HIF-1) is related to resistance to chemotherapy and radiotherapy targeting HIF-1 may help improve anti-tumor treatment.

Apoptosis and autoimmunity are two phenomena, which share essential similarities in destroying cells. However, apoptosis is a normal process, while autoimmunity is not. Apoptosis could, at least in some conditions, cause autoimmunity. In the whole body, more than 50 billion cells die each day due to apoptosis. Apoptosis plays a role also in the negative selection of thymocytes and lymphocytes. It has been estimated that as many as 95% of T lymphocytes produced are eliminated by apoptosis, presumably because they recognize self-antigens.

The "decision" to activate a suicidal process is made based on intrinsic or extrinsic apoptotic messages. Intrinsic inducers come from mitochondria or the nucleus. Extrinsic inducers are ligands and cytokines for death receptors on the cells' surface.

But apoptosis can be induced also by a lack of signal molecules produced by other cells, which normally work in combinations to regulate the behavior of the cells. Activation of naïve T cells by antigens results in a massive expansion of antigen-specific T cells. The majority of T cells, CD4+CD25+ suppressor T cells, die due to apoptosis, while a small number survive and differentiate into memory cells. This cell density is crucial for our understanding of how autoimmunity is evaded. Some auto-reactive cells may escape natural apoptosis and represent continuous treatment with a potential autoimmune response.

Apoptotic cells do not disappear after phagocytosis. Ingredients of apoptotic cells indeed survive the intra-cellular processing and are recycled to the membrane of phagocytes. Massive apoptosis overloading the phagocytic capacity may trigger an autoimmune reaction through the presentation of nucleosomes to the immune system. Apoptosis has traditionally been thought of as a non-inflammatory process which does not induce an immune response.

However, recent studies indicate that apoptotic cells can be involved in the auto-immune process. They can display auto-reactive antigens in their surface blebs, can activate dendritic cells and they can induce the formation of auto-antibodies.

In systemic auto-immune diseases, especially in systemic lupus erythematosus (SLE), auto-antibodies are found against a palette of intra-cellular antigens. It has been suggested that humoral response may be driven by the products of apoptotic cells with the central importance of the nucleosomes as potential auto-antigens.

Auto-reactive of B and T cells, escaping from natural apoptosis might represent an additional necessary condition. The increased activity of CD3+, CD4, and Th1 "helper" dominant cells results in elevated serum levels of IL-Beta, IL2, TNF-alpha and IFNɣ.

Caspase activity in apoptotic cells may lead to the presentation of cryptic epitopes or neo-epitopes to which the immune system is not tolerant: phosphorylated, citrullinated, and acetylated, which may lead to increased immunity. Direct access to the cytosol of antigens derived from the processing of engulfed apoptotic cells has been reported in activated phagocytes. It has been proposed that possible increased sensitivity to CD95 provokes apoptosis:

- epithelial cell death is mediated by CD95 expressing lamina propria lymphocytes (ulcerative colitis);
- high levels of soluble CD95, found in rheumatoid arthritis (RA), contributes to the inhibition of synoviocyte and inflammatory cell apoptosis;
- inadequate apoptosis due to defective CD95 may promote extending survival of synoviocytes and additionally, their responsiveness to CD95L is decreased by TGF beta interleukin-1 beta, and TNF-alfa;
- expressed CD95 and its ligand cause apoptotic cell death by paracrine or autocrine mechanism and during inflammation interleukin-1 beta and interferon-1 alpha induce massive CD upregulation.

Studies from the last 10 years have suggested that the neonatal wave of beta cell apoptosis might provide auto-antigens necessary for triggering beta cells' direct auto-immunity in diabetes mellitus. In the animal models increased TNF-alpha-mediated apoptosis has been found, which has been explained by selective nitric oxide-mediated upregulation of the functional CD95 molecule which is subsequently killed by CD95L-producing T cells. IAPs inhibit apoptosis in part through the prevention of activation of some pro-caspases.

Beside HIF-1, a different mechanism of inducting apoptosis is by stabilization of wild-type p53 by HIF-1. If the cells have already a p53 mutation, this happens via hypoxia-induced apoptosis through the induction of the expression of anti-apoptotic protein IAP-2. Many studies have focused on the role of HIF-1 in angiogenesis, and it is clear that the function of HIF-1 in the regulation of apoptosis by hypoxia deserves more attention.

A better understanding of the regulation of apoptosis by hypoxia in solid tumors may enhance insight into tumor behavior and the effect of hypoxia on anti-tumor treatment. One of the key players in cancer development is the p53 gene, a tumor suppressor. Protein p53 is a transcription factor encoded by a gene whose disruption is associated with approximately 50 to 55 per cent of human cancers.

The p53 protein acts as a checkpoint in the cell cycle, either preventing or initiating programmed cell death (apoptosis). Since cancer is the unchecked proliferation of cells, disruption of p53 activation allows diseased cells to multiply or trigger a tumor's progression from benign to malignant.

Functional profiling of this protein allows precise characterization of mechanisms of its regulation. In normal conditions, the genome is tightly guarded by the p53 gene. When DNA damage is encountered p53 expression is up-regulated and p53 protein accumulates in the cells through increased transcriptional activity. The cell cycle may undergo arrest or undergo apoptosis in response to these conditions.

1.3. Measuring and interpreting the selectivity of protein kinase inhibitors

Protein kinase inhibitors are a well-established class of clinically useful drugs, particularly for the treatment of cancer. Achieving inhibitor selectivity for particular protein kinases often remains a significant challenge in the development of new small molecules as drugs or as tools for chemical biology research.

This review summarizes the methodologies available for measuring kinase inhibitor selectivity, both in vitro and in cells. The interpretation of kinase inhibitor selectivity data is discussed, particularly concerning the structural biology of the protein targets. Measurement and prediction of kinase inhibitor selectivity will be important for the development of new multi-targeted kinase inhibitors.

Protein kinases catalyze the transfer of the terminal phosphate group from adenosine triphosphate (ATP) to the hydroxyl group of a serine, threonine or tyrosine residue of the kinase itself or another protein substrate. The addition of phosphate groups can create new recognition sites for other proteins to bind to or may alter the conformation of the phosphorylated protein and change its activation state or function.

Protein kinase enzymes play pivotal roles in signal transduction, the transfer of a signal from the cell membrane to the nucleus and cell-cycle control. Some tyrosine kinase (TK) domains are part of receptor tyrosine kinases; proteins that span the cell membrane and are involved in transducing growth and survival signals into the cell in response to external growth factor stimuli.

Because of their importance in cell-cycle progression, cell growth, apoptosis and metabolism amongst other processes, modulating protein kinase activity has the potential to treat diseases, particularly cancer and other signals. Kinases involved in inter- and intracellular signaling are frequently found to be mutated in cancer cells, leading to constitutive over or under-activation. This abnormal signaling can result in uncontrolled cell growth and proliferation and so by specifically targeting the signal transduction pathways in which oncogenic kinases are involved it may be possible to prevent aberrant cell growth and other behaviors.

1.4. Apoptotic Regulations

Initially, apoptosis can coexist within a single cell. However, all of these experiments were performed with antibodies to CD95, and it is now known that their action is not identical to that of FasL. It is therefore imperative to reevaluate CD95 signaling in experiments by using the physiological ligand (Strasser and O'Connor, 1998).

The role of adaptor proteins in regulating pro- and anti-apoptotic signaling pathways induced by the activation of death receptors is very critical. Adaptor proteins, such as FADD (Fas-associated death domain) and TRADD (TNF receptor-associated death domain), are recruited to ligand-activated, oligomerized death receptors to mediate apoptotic signaling pathways.

Apoptotic adaptors associate through domains homologous with the death domain (a cytoplasmic portion) of death receptors. After association with a ligand-bound receptor, the adaptor proteins recruit procaspase-8 or procaspase-10 with the help of their death effector domains (DEDs) to form the death-inducing signaling complex (DISC), which initiates a caspase signaling cascade and consequently induces apoptosis. Other adaptor proteins initiate apoptotic signaling through distinct mechanisms.

CRADD/RAIDD (caspase and RIP adaptor with a death domain) and PIDD (p53-induced protein with a death domain) are adaptor proteins that associate with procaspase-2 to form the PID Dosome following DNA damage and lead to the cleavage and activation of caspase-2.

RIP1 (receptor inter-acting protein-1) and DAXX (Fas death domain-associated-xx) are adaptor proteins with a death domain, which are recruited to TNFR1 or Fas, respectively, upon ligand binding. Apart from promoting apoptosis, ligand binding to DR3 (TWEAK) or TNFR1 can cause the intracellular adaptor proteins to recruit TNF receptor-associated factor-2 (TR, which subsequently leads to the activation of NF-beta.

B is responsible for inducing the expression of anti-apoptotic genes such as FLIP and Bcl-2. Cell survival and death are well-regulated phenomena that are partly decided by the type of ligands that bind to the surface receptors. Death ligands like Fas and TNF bind to their receptors to promote apoptosis via caspase activation. Anti-apoptotic ligands-like growth factors bind to the cell surface receptors to activate intracellular kinases (e.g., PI3K, Akt/PKB, PKA, PKC, MEKs, ERK1/2, p90RSK, and p38 MAPK), which promotes cell survival. The activation of PI3K leads to the phosphorylation and activation of Akt/PKB, which results in the Akt-dependent phosphorylation and the inhibition of pro-apoptotic molecules such as Bad, GSK-3, and caspase-9.

Kinases were not necessarily considered good drug targets when the study of pharmacological inhibitors began. As the cellular concentrations of ATP are high, typically 1–5 mM, it was thought that a high concentration of the drug would be needed for efficacy with ATP-competitive inhibitors, bringing potential toxicity problems.

Another concern was that a large number of protein kinase enzymes, all sharing a common cofactor and similar three-dimensional structure of the catalytic site, might confound attempts at developing adequately selective drugs. As small-molecule inhibitors of protein kinases are now well-established as clinically useful drugs, particularly for the treatment of cancer it can be seen that these problems have not proven to be insurmountable.

Kinase-dependent activation of transcription factors such as CREB and Stat family members (Mad, Max, and c-Myc) promotes cell survival by inducing the expression of Bcl-2 and Bcl-xL. Max heterodimers are associated with transcriptional repression and cellular differentiation, while Myc-Max heterodimers are associated with transcriptional activation and cellular proliferation.

The Stat family of transcription factors is activated by intracellular kinases such as Jak (1 to 3), TyK2, and Src. In normal cells, Bad is phosphorylated and sequestered in the cytoplasm by the adapter protein. The increase of cytoplasmic levels of free Bad causes anti-apoptotic proteins Bcl-2 and Bcl-xL to bind with Bad, and consequently release Bax and BAK. Bax and BAK initiate the release of cytochrome c from the mitochondrial membrane.

There are currently eight approved kinase inhibitors in use in clinics which bind in the ATP binding site. These target a range of protein kinases, although tyrosine kinase inhibitors dominate, and are used to treat a variety of cancer types. Considerable medicinal chemistry effort has been directed towards kinase targets, and this has increased over the last decade.

However, routinely achieving drug selectivity for particular protein kinases remains a significant challenge. Defining, measuring and engineering inhibitor selectivity has become a critical activity, both in the development of new drugs and in the application of the inhibitors as tools for chemical biology research.

While research initially focused largely on finding mono-specific or very highly selective kinase inhibitors for therapeutic use, clinical experience and a growing understanding of kinase biology show that compounds with a broader pattern of inhibitory activities can be effective. Nevertheless, such compounds still need to be selective inhibitors of a well-defined set of kinase enzymes. This review includes the methodologies available for measuring and interpreting kinase inhibitor selectivity, both in vitro and in cells, and describes how these data could be applied to the development of new multi-targeted kinase inhibitors.

1.5. Protein kinase structure and selective inhibition

There are more than 500 protein kinases coded for in the human genome. These have been grouped into families based on their similarity in the amino acid sequence of the catalytic domain. They comprise of two main domains, N- and C-terminal domains. The cofactor ATP binds to the backbone of the linker region between the domains, also called the hinge region.

Many protein kinases have a catalytically active and inactive state. In the inactive state, it has been noted that the structures of the catalytic domain are more diverse. When the kinase is activated, often by phosphorylation on the activation loop, it adopts a shape which can bind ATP with the binding site open and the necessary residues correctly orientated to carry out the phosphate transfer.

The DFG motif is a conserved sequence of three amino acids, aspartate (D), phenylalanine (F) and glycine (G) which forms part of the ATP binding site in the active kinase. A further conserved aspartate residue in the catalytic domain may aid in the phosphate transfer by deprotonating the substrate hydroxyl group enabling it to more readily attack and remove the terminal phosphate from ATP.

The catalytic domain in the active conformation has been well characterized and an early pharmacophore model built to describe the ATP binding site has been successfully used for rational drug development. Although the ATP binding site is quite conserved, the adjacent areas not occupied by ATP itself are more variable, particularly the interior hydrophobic pocket and the solvent-exposed specificity surface. Thus, by making inhibitors that bind to these areas, more specificity may be obtained. An example of a kinase inhibitor which targets these areas is dasatinib.

Kinase inhibitors can target the active or inactive conformations of the enzyme and may be directed towards the ATP-binding site, the substrate-binding site or allosteric sites. There are benefits and drawbacks associated with each of these strategies. One benefit of targeting the active state is that the ATP binding site is well-characterized and many X-ray crystal structures are available which can aid in the design of new inhibitors. The availability of X-ray crystal structures and the rigidity and conserved structure of the ATP binding domain enable rational inhibitor design using docking and virtual screening methods. On the other hand, as the ATP binding site structure of active kinases is so conserved throughout the class, it may be more difficult to gain specificity for a particular kinase.

For inactive kinases, although specificity is possibly easier to obtain as a greater diversity of protein conformations is possible, less is known about these conformations, so designing inhibitors becomes more of a challenge. Nevertheless, general pharmacophore models of inhibitor binding to inactive kinases have been developed and there are several kinase inhibitors which target this inactive state, such as Sorafenib.

A key feature of the inactive kinase protein structure is the opening up of an extended hydrophobic pocket as the activation loop adopts the 'DFG-out' position and disrupts the arrangement of the catalytic residues. A version of this model has been applied to rationally design inhibitors that bind and stabilize the inactive conformation of protein kinases.

Through in-silico modelling of the catalytic site in the inactive conformation, it was proposed and subsequently confirmed that the incorporation of a large, lipophilic trifluoromethyl-benzamide group at certain positions of known active conformation inhibitors could transform them into inhibitors binding the inactive kinase conformation.

One of the main problems that have emerged with targeting inactive kinases is the susceptibility to loss of inhibitor activity due to mutation. In an active kinase, a mutation preventing the binding of an inhibitor will often abolish kinase activity (by preventing ATP binding or catalysis) and, therefore, cannot be tolerated. In contrast, inhibitors bind to different residues in the inactive kinase, mutations of which are often less important for ATP binding and catalytic activity and are, therefore, better tolerated. The mutated form remains a functional enzyme, but one to which the inhibitor can no longer bind.

This has been demonstrated in the case of imatinib where mutations in the catalytic domain of the Abl kinase can lead to imatinib-resistant chronic myeloid leukemia (CML), a disease that is driven by the Bcr-Abl oncogene. There are several possible mutations leading to imatinib resistance. Many patients have only one single point mutation preventing drug binding and leading to relapse. One of these mutations is T315I, which is a gatekeeper mutation preventing inhibitor access to the hydrophobic pocket and removing one of the crucial hydrogen bond donors used for imatinib binding.

Targeting sites on protein kinases other than the ATP binding site is a means of circumventing the possible problem of competing with high cellular ATP levels. Substrate or protein partner binding sites can be targeted, although these often involve protein–protein interactions over large surface areas which are difficult to inhibit with small molecules.

Some kinases also contain allosteric sites, which are pockets remote from the ATP binding site into which inhibitors can bind, altering the overall conformation of the catalytic domain and inhibiting the enzyme. These may be more easily targeted by small molecules than substrate binding sites as they do not involve disrupting protein-protein interactions, but not all kinases have a known suitable allosteric site.

Nonetheless, because allosteric inhibitors may avoid many of the problems associated with ATP-competitive kinase inhibition, there is a lot of interest in finding them. In particular, allosteric sites offer the possibility of very highly selective or mono-specific inhibitor profiles. Proof of concept for this approach has been demonstrated by the discovery of potent and highly selective inhibitors of Akt that bind to an allosteric site located between the kinase domain. The inhibitors are selective not only for Akt over other very closely homologous enzymes in the AGC subfamily but can also differentiate between the three isoforms of the kinase.

1.6. The Importance of determining selectivity

Many kinase inhibitors already in clinical use to treat cancer are not selective for one particular kinase but target several. This can not only bring efficacy benefits where the other kinases inhibited are also oncogenic or involved in driving cellular malignancy, but it also potentially increases the risk of unwanted effects and toxicity.

Of course, side effects can be due to interaction with other enzymes, ion channels or receptors as well as the inhibition of other kinases. It is important, therefore, to know what biomolecular targets drugs can interact with and what enzymes they can inhibit.

In recent years, it has been pointed out that there can be benefits to having selective but non-specific inhibitors as drugs. One is that a spectrum of kinase inhibitory activities can lead to approved drugs being effective for treating diseases other than those initially targeted. Imatinib, which was developed and first used in the clinic as a Bcr-Abl inhibitor to treat CML, was subsequently shown to have clinical efficacy against gastrointestinal stromal tumors (GIST) due to its inhibition of c-Kit. Dasatinib is a dual Src/Abl kinase inhibitor which has been shown to be effective against a range of tumor types.

The compound also inhibits PDGFR and its activity has been attributed to its multi-targeting of tyrosine kinases. Nilotinib was developed as a second-generation agent for CML in response to the observation of drug resistance in imatinib-treated patients. The compound has activity against almost all known mutant forms of Bcr-Abl.

It also inhibits c-Kit and PDGFR, which are oncogenic targets in several tumors such as GIST. Again, this is a multi-tyrosine kinase inhibitor and this multitargeting is a potential benefit in the clinic as the same drug can be used to treat a variety of cancers and overcome some resistance problems. The clinically approved agents sunitinib and sorafenib are also inhibitors of multiple tyrosine kinases.

Lapatinib is a dual inhibitor of Erb-2 and EGFR. Overexpression of either Erb-2 or EGFR receptor tyrosine kinases is implicated in a range of cancer types, and overexpression of both in ovarian cancer leads to a poorer prognosis. Thus, a dual inhibitor of these targets is a potentially desirable profile. Tumor growth arrest by lapatinib was shown in human tumor xenografts, and the compound was approved for the treatment of breast cancer in 2007. Other groups have also investigated dual Erb-2/EGFR inhibitors.

The concept of targeting a defined range of protein kinases is now increasingly apparent in reports of preclinical drug discovery, and there are many examples of dual inhibitors in the recent literature. These sometimes target kinases in the same kinone sub-family but can also be directed against two functionally related kinases from different kinone sub-families, as in the example of dual inhibitors of Chk1 and Wee1 which are both involved in the G2 cell cycle checkpoint. Studies by several groups have shown that dual inhibitors of the protein kinase mTOR and the lipid kinase phosphoinositide 3-kinase (PI3K) may have better efficacy in certain cancer types compared to more specific inhibitors. The design of dual mTOR/PI3K inhibitors addresses a potential limitation of the mTOR inhibitor rapamycin.

Inhibition of mTOR by rapamycin and analogues causes a compensatory upregulation of the upstream signaling kinase Akt due to the release of a negative feedback loop. By simultaneously inhibiting the pathway at two points upstream and downstream of Akt, this can be overcome. Although PI3K is a lipid kinase rather than a protein kinase, it is also involved in oncogenic pathways and contains an ATP binding site similar to the protein kinases, allowing it to be inhibited with similar compounds.

Thus, for drugs, inhibiting more than one kinase may be beneficial as long as these are therapeutically useful targets; for example, in overactive pathways in cancer. However, knowledge and control of selectivity are still important as off-target kinase inhibition can also cause toxicity.

For example, the inhibition of AMPK and some tyrosine kinases has been linked with cardiotoxicity through the sensitivity of cardiomyocytes to the disruption of cellular energy production. Some TKs have been implicated, including Abl, which has an identical kinase domain to the fusion protein Bcr-Abl that is the primary target of imatinib in CML.

To overcome the cardiotoxicity of Abl inhibitors, a structure-based approach involved adding a methyl group to imatinib to prevent binding to Abl and Bcr-Abl while retaining other TK inhibitory activities. As the anti-tumor activity of imatinib in some tumor types is due to inhibition of kinases other than Bcr-Abl, notably c-KIT in GIST, new compounds without the Abl activity still have the potential for efficacy in these diseases.

Measuring the selectivity and understanding the common features of the binding of unselective kinase inhibitors may help in the design of less promiscuous inhibitors or at least allow hits from screening libraries to be priority to select scaffolds with greater potential for specificity.

This approach may also be useful in the design of multi-targeted inhibitors with controlled selectivity profiles by defining chemical classes that have a consistent bias towards a clear selectivity pattern. Moreover, there is an argument that inhibitors that are to be used as pharmacological tools to understand biological systems need to be as selective as possible.

If a compound inhibits multiple kinases, its biological effects may not be easily attributable to specific targets, making their function more difficult to elucidate. At the least, knowledge of the kinone inhibitory profile of tool inhibitors is an important part of assessing their fitness for use as pharmacological probes.

Recently, it has become common practice that the inhibitory activity of new inhibitors is measured against a panel of protein kinases. Many groups choose to concentrate on kinases in the same family and, therefore, with similar ATP binding site sequences. Another approach is to measure the effects on functionally related kinases, as inhibiting more than one enzyme on the same signaling pathway may make a significant difference to the action of a drug.

Many others screen against a wide variety of kinases to get an overall view of selectivity. This wider kinone profiling was not always applied, and it is now being found that compounds which were thought to be mono-specific or highly selective inhibitors are not as selective as first believed.

1.7. Measuring the inhibition of substrate phosphorylation

There are many ways of measuring kinase activity, and these are reviewed regularly in the literature. They have varying advantages and disadvantages, and different techniques may be more suitable for a particular kinase or compound set. One of the first and most widely used assays involves radio-labelled [32P]- or [33P]-ATP. This allows the direct detection of phosphorylation of a substrate peptide or protein by a kinase of interest.

While measuring the incorporation of radioactive phosphate directly provides few sources of interference in the assay, there can be associated limits on the amount of radioactivity that can be used and there is the need for specialist disposal. As a result, non-radioactive methods are also widely used, such as fluorescent and luminescent endpoint assays, although these may have their disadvantages. One is the fact that a significant number of organic compounds are colored or fluorescent and can interfere with the results.

Another is that many of the assay formats require specific phosphor-peptide antibodies to detect the conversion of a substrate. Although there is an available phosphor-tyrosine antibody which can be used for any phosphorylated tyrosine residue, it has proven much harder to find general antibodies recognize serine/threonine phosphorylation, and separate antibodies are needed for each peptide used. Thus, significant time and effort may be needed to develop antibody reagents for such assays.

Another more recently introduced way of directly measuring substrate phosphorylation is by using mobility shift assays. This method involves the electrophoretic separation of phosphorylated and non-phosphorylated short peptides based on their charge. The peptides are tagged with a fluorescent marker for detection. These assays can be carried out in a 384-well plate format and have been used in high-throughput to screen 32,000 compounds as well as for screening inhibitors against kinase panels.

Electrophoretic separation assays have been compared with a radiometric assay format and found to be more reliable. In our research, we have used this technology to profile both potent kinase inhibitors and also to investigate the selectivity patterns of low molecular weight fragments (MW < 200) tested at high compound concentrations.

1.8. Measuring the binding affinity of kinase inhibitors

As well as biochemical assays to measure phosphorylation, competitive binding assays can be used to measure the dissociation constant (*K*d) of an inhibitor-kinase complex. These generally involve an ATP mimic or competitive inhibitor which is used as a probe and is in competition with the inhibitor of interest. The *K*d can then be calculated from the amount of probe bound to the kinase compared to a control.

The probes are often attached to fluorescent markers to detect them. In one case, tauro-sporine or close analogues have been used as probes without additional markers, since excitation of the compounds at 296 nm results in emission at 378 and 396 nm. The fluorescence is enhanced on binding to the ATP binding site of the kinase, allowing a comparison between binding in wells with and without inhibitor.

Although this study looked at screening compounds against a single kinase, Src, one could see the possibility of using the same probes to profile an inhibitor against several kinases. There was interference from some of the compounds tested, as they fluoresced at the same wavelength as the tauro-sporine analogues.

The pharmaceutical company Ambit Biosciences has described a binding assay to determine the *K*d of up to 38 kinase inhibitors against 317 protein kinases using an immobilized 'probe' ligand which binds to the ATP binding site. Human kinases were expressed and fused to a bacteriophage and were incubated with the immobilized probe and the inhibitor of interest. The kinases are bound to either the probe ligand or the unbound test compound.

After a wash step, the amount of kinase bound to the immobilized probe was determined by quantifying the immobilized bacteriophage using plaque assays or a quantitative polymerase chain reaction. The binding constants obtained by these methods agreed well with both literature *K*d values and the *K*i or IC50 values from phosphorylation inhibition assays.

Another way of assessing inhibitor binding to kinases is to use affinity chromatography. This technique has been applied to look for inhibitors of specific kinases by immobilizing the enzyme onto beads and running a solution of small molecules through it. By coupling this to mass spectrometry, the ligands which bind can be identified. The same approach can also be used to look at inhibitor selectivity by immobilizing the inhibitor and running proteins or cell lysates through the functionalized resin.

Bound proteins are eluted from the affinity resin, purified by gel electrophoresis or western blotting and identified by immunoblotting, protein staining or mass spectrometry. One limitation of immunoblot detection is that the choice of antibodies determines which protein targets are detected. This approach was used to investigate the interactions of sunitinib, and several new targets were found. Affinity chromatography of the cell lysate, followed by electrophoretic separation of the captured proteins and their identification by mass spectrometry, was employed to determine the cellular targets of the EGFR inhibitor gefitinib.

Immobilized roscovitine was also studied in this way, and binding to pyridoxal kinase was discovered in addition to the expected targets. Although effective for identifying targets and detecting the pattern of proteins an inhibitor can bind, one limitation with affinity chromatography of this type is the inability to immediately quantify the relative strength of binding when cell lysates are used.

The relative abundance of each kinase present may affect how much of each target is captured as well as the strength of the binding interaction. The heterogenous affinity-based methods measure apparent *K*d values and may not always reflect the *K*d values in the solution. However, binding affinities for different compounds can be compared; thus, the relative selective of multiple inhibitors can be evaluated.

A biophysical method for assessing inhibitor binding to kinases uses thermal shift data. This compares the difference in the melting temperature (*T*m), the temperature at which the protein unfolds, in the presence or absence of the inhibitor. The degree of denaturation is measured using a peptide-binding dye. Tight binding inhibitors stabilize the peptide in its folded conformation, raising the *T*m. Using this technique, the selectivity of 156 kinase inhibitors against a panel of 60 serine/threonine kinases was assessed.

1.9. Microarrays, miniaturization and proteomic methods

To improve efficiency, some of the assays described above have been miniaturized so they can be run on 1536- and 3456-well plates and used for high-throughput screening and kinase profiling. The main advantages of these techniques are the reduction in quantity of materials needed and the increased amount of information obtained. Even more, screening can be carried out at once by using microarrays, of which there are numerous types involving the immobilization of peptides, proteins or ligands.

Several types of surfaces are used, including glass, nano well and 3D-surface structures. Immobilization can be carried out by using high affinity interactions such as streptavidin–biotin, surface adsorption or covalent interactions.

These and other considerations are covered in detailed reviews. To determine the inhibitory profile of single compounds against multiple kinases, the most useful approach would be to use a protein array. However, there are problems associated with immobilizing enzymes on chips as they may denature easily and lose activity.

The orientation of the bound proteins concerning the surface is also important so that none of the binding sites is blocked or altered by the fixation. Different post-translational modifications may lead to proteins with varying activity so it may be necessary to make several forms of the protein to investigate the activity fully.

These issues have meant that interactions are not always seen where expected, and it is advisable to use multiple techniques to confirm the activities observed.

A schematic shows the measurement of inhibitor selectivity using: a) a microarray of immobilized kinases; b) a microarray of immobilized substrate peptides; c) the yeast-3-hybrid proteomic method (*AD* activation domain, *BD* DNA binding domain); and d) tandem MS detection.

Peptide substrates or test compounds have been simply arrayed on glass slides and other assay components sprayed onto the chips, avoiding some of the problems associated with immobilized protein microarrays. Potential inhibitors can be arrayed and the enzymes and fluorescence-tagged substrates subsequently sprayed onto the chip. Various detection methods are used, for example, in some cases the substrates fluoresce where they have been phosphorylated by the enzyme of interest.

This method has been adapted for kinase inhibitor profiling using an immobilized tyrosine-containing peptide substrate on streptavidin slides. The test inhibitors were arrayed on top of the peptide and the appropriate kinase was sprayed onto the slide.

As the phosphorylated peptide was immobilized, wash steps could be included in the assay, and antibodies were used to recognize the phosphorylated tyrosine followed by a second fluorescent-labelled antibody. The technique was demonstrated using four kinases, but an extension to a wider panel could be envisaged.

Patterns of kinase inhibitor selectivity can emerge as part of proteome-wide screening as well as from dedicated kinase profiling. For example, the yeast three-hybrid approach has generated new insight into the selectivity of cyclin-dependent kinase (CDK) inhibitors. In this case, the compounds were covalently attached to the dihydrofolate reductase (DHFR) inhibitor methotrexate via a PEG spacer. A DNA-binding domain was fused to DHFR, and cDNA libraries were used to express large numbers of proteins fused to an activation domain.

On binding to a protein, the CDK inhibitor-methotrexate conjugate brought the DNA binding and activation domains into proximity, inducing detectable expression of the reporter gene. Interestingly, it was noted that the only targets identified for the CDK inhibitor purvalanol B were kinases, giving a wider context to the selectivity pattern. A detailed review of proteomic methods is available.

The development of tandem mass spectrometry (MS) techniques for the parallel quantification of the protein content of cell lysates has provided a set of analytical methods that can be adapted for kinase inhibitor selectivity profiling. In particular, isobaric tags for relative and absolute quantification (iTRAQ) reporter reagents for the labelling of peptides before MS analysis have been used in conjunction with affinity chromatography to assess the affinities of kinase inhibitors for endogenously expressed protein kinases. Affinity chromatography was conducted with beads coated with a mixed population of seven, non-selective kinase inhibitors.

This matrix was shown to effectively capture 183 distinct protein kinases from K562 tumor cell lysates.

Three Bcr-Abl kinase inhibitors (dasatinib, imatinib and bosutinib) were added at different concentrations to the cell lysate before affinity chromatography. The bound proteins from the control and drug-treated affinity purifications were digested with trypsin and labelled with different iTRAQ reagents. The combined proteolytic digest was analyzed by tandem MS to detect the variation in the pull-down of kinases between the drug-treated and control samples.

Dose–response curves for inhibitor binding to approximately 150 kinases were generated from these data. An alternative approach used competition of inhibitors with specific chemical-labelling probes for the kinase ATP binding site to derive modified peptides for MS analysis. ADP or ATP bearing reactive terminal acyl groups, further substituted with a biotin group, were incubated with cell lysates.

The positions of the reactive acyl group of the probes, close to conserved lysine at least one of which is present in almost all protein kinases, led to specific transfer of the biotinylated tag to the kinase. Approximately 75–80% of the enzymes in the human, mouse and dog kinone were effectively labelled by these acyl phosphates.

After trypsin digestion, biotinylated fragment peptides were captured using streptavidin-coated beads. Identification and quantification of the labelled peptides were achieved by tandem MS. As with the iTRAQ method discussed above, comparison of the labelled peptides obtained with inhibitor-treated and control lysates allowed the construction of dose–response curves to determine inhibitor-binding affinities.

The selectivity profile obtained for tauro-sporine using this technique contained anomalies when compared to that produced by conventional biochemical or phage-display methods, with the affinities for some kinases being two orders of magnitude less than the values reported from phage-display.

Differences in activation state between recombinant enzymes and endogenously expressed full-length kinases may account for some of this variation. It has been suggested that the kinases from lysates are likely to preserve important post-translational modifications and protein partner interactions that will modify the activity of the enzymes, although the significance and biological relevance of these differences remains unclear.

1.10. Kinase selectivity profiling in cells

The kinases are expressed with a mutation in the ATP binding site that renders them sensitive to an ATP analogue that is usually too large to bind and, thus, does not affect the wild-type enzymes.

The effect of specific inhibition of each kinase in the cell can, therefore, be observed separately on gene expression microarrays. The cellular targets of a novel small molecule inhibitor in cells can be inferred from comparing the observed transcription profile with that due to inhibition of the single mutated kinases.

It has been pointed out that although many targets can be identified in theory, the data may be difficult to interpret when many kinases are being inhibited, and the technique may be most useful for relatively selective compounds.

An example where kinase selectivity in vitro was shown to translate to the cellular context was seen with ATP-competitive inhibitors of Ak. Inhibitor **1** was 30-fold selective for PKA over Akt in a biochemical assay. The selectivity in cells was probed by monitoring the specific, and different, downstream effects of inhibiting these enzymes, where a similar selectivity for inhibition of PKA over Akt was observed. It is noteworthy that the phenotypic effects of inhibition of different kinases may require different assay formats for detection. Another example illustrates the case where in vitro selectivity does not translate to an equivalent cellular selectivity.

A differential in sensitivity was observed in cells with an inhibitor with no intrinsic difference in affinity for two kinases. The CDK1 inhibitor BMI-1026 showed similar potency against CDK1 and CDK2 (8 and 2 nM IC50s, respectively) in biochemical assays. However, on treating cells at 40 nM, only CDK1-dependent cell cycle arrest could be seen. Although at higher concentrations, CDK2-dependent cell cycle arrest was also observed, the results suggested that this cell line was inherently more sensitive to CDK1 inhibition.

Compound **2** also illustrates a case where in vitro selectivity and cellular selectivity do not correlate. Here, biochemical assays showed **2** to be 20-fold selective for inhibition of CDK1 over Pho85. However, in cells, only an effect due to the inhibition of Pho85 was observed. This was attributed to a differential sensitivity between CDK1 and Pho85 to inhibition.

Although the intrinsic affinities of inhibitors may often be reflected in their phenotypes in cellular systems, it can also be seen that the selectivity of compounds may change between biochemical and cellular assays. This could result from differences in *K*m, ATP of the kinases as discussed above, differences in the abundance of the kinases in the cells or differences in the sensitivity of the cells to inhibition of particular pathways.

This last factor may enhance the cellular selectivity of a small molecule if the intrinsic inhibitor affinities and the inherent pathway sensitivities align. In interpreting the in vitro data, it might be necessary to attain a large threshold, perhaps 1,000- or 10,000-fold selective, to be assured that the in vitro kinase selectivity will dominate in the cellular context. For this reason, it is important to test compounds in relevant cellular assays, although biochemical kinase profiling remains a useful and tractable first step in looking broadly at the potential activities and selectivity of compounds across the kinone.

There are additional factors to consider when moving from cells into living organisms that will affect the translation of a biochemical or cellular selectivity profile into physiological effects. Differential distribution of the drug to tissues, coupled with the possibility of differential expression of kinases between tissues, will modify the significance of the apparent selectivity profile.

For example, although Sorafenib was rationally designed to inhibit Raf kinase, and has good potency on this target, it also inhibits several other kinases, including many of the growth factor receptor tyrosine kinases which are involved in angiogenesis. Some of the clinical efficacy of the drug can be ascribed to the inhibition of these kinases in normal blood vessel epithelia, preventing the development of new blood supply from surrounding tissue to the tumor, in addition to the inhibition of enzymes in the tumor cells themselves.

1.11. Predicting specificity and selectivity of kinases

As kinase screening becomes more common, there have been many different approaches suggested for improving specificity and selectivity. The multiplexing of high-throughput screening, so that compound library is screened against many kinases in parallel, has been proposed as an alternative to single target-centric discovery of new starting points for drug discovery.

The availability of the miniaturized and high-throughput methods discussed earlier makes this a realistic proposition. Researchers at Vertex proposed a pharmacophore for kinase 'frequent hitters' which involved comparing the structures of inhibitors shown to have activity on a set of five protein kinases.

The pharmacophore is suggested as a filter to remove promiscuous inhibitors from screening libraries or as the start-point for new medicinal chemistry projects.

The '2–0 rule' has also been introduced to help identify likely kinase inhibitors. The '2–0 rule' states that a compound is likely to have kinase activity if it contains: two or more heteroaromatic nitrogen, one or more heteroaromatic NH groups, one or more aniline and one or more nitriles. This can be used to avoid common motifs or to identify potential new kinase inhibitor scaffolds, with a 5-fold enrichment in the probability of identifying a kinase inhibitor motif found in the test set.

X-ray crystallography data are particularly valuable in examining in detail the differences between the ATP binding site in active and inactive conformations or between different kinases.

Where co-crystal structures of kinases with various compounds from the same chemical class are solved, these can be used to improve the specificity by structure-based design. For example, a single amino acid difference was found to be the major determinant of selectivity for Akt versus PKA in two series of Akt inhibitors.

A library of fragments has been screened against 10 kinases to investigate if these very low molecular weight compounds show intrinsic specificity. It was noted that even the fragments had some pattern of selectivity, inhibiting only one or two kinases at test concentrations of 1–100 μM.

The fragments were elaborated using X-ray crystallography as one of the main tools to make decisions about where to add new functional groups. When the fragments grew to these larger inhibitors, sub-micromolar inhibitory activity was seen on all kinases tested, but most compounds only hit one or two kinases very potently.

Another way of predicting possible kinase targets for compounds are to compare the amino acid sequences of the catalytic sites or particular residues within them. One very important residue in distinguishing between inhibitors is the gatekeeper residue. This is found at the entrance to the hydrophobic pocket. If the residue is small, the hydrophobic pocket is accessible, but larger gatekeepers block this pocket and stop inhibitors which occupy this part of the catalytic domain from binding. For example, compound **3** inhibited five out of seven tyrosine kinases tested.

The five kinases in which compound **3** was active all had a serine gatekeeper. The two kinases which were not inhibited at the test concentration had larger methionine and leucine gatekeeper residues, suggesting a possible rationale for the observed selectivity.

In addition to looking at the kinase structure, the available structure–activity data can be used to look for kinases that are inhibited by structurally similar compounds. The distances between key features in the kinases and possible inhibitors can be compared. These and other similar in silico methods are part of what has become known as 'chemogenomics', the discovery and description of all possible drugs for all possible drug targets. Several groups have looked at using a selection of these methods to predict which kinases will be inhibited by a compound.

Some good correlations between calculated and experimental data were found as long as the compounds used for training were sufficiently structurally similar to those subsequently tested. An analysis of the binding modes of many kinase inhibitors, as determined by X-ray crystallography, has suggested simple rules of thumb for predicting the orientation of typical ATP-competitive inhibitor scaffolds within the binding site [49].

Chapter Ten

New Cancer Therapies for Hematological Malignant Diseases with p53 Mutant Genes and Isoform p53 Proteins

Some studies have shown that patients with cancer make antibodies against p53 proteins, but the frequency and magnitude of this response are still under debate [56]. The most advanced work used a long synthetic peptide mixture derived from p53 (p53-SLP; ISA Pharmaceuticals, Bilthoven, the Netherlands). The vaccine is delivered in the adjuvant setting and induces T helper-type cells. About one-third of cases (30%) did not have recurrent chromosomal mutations, suggesting a high degree of genetic heterogeneity and no clear mutational drivers of CLL. Consistent with a role in disease initiation, global DNA hypomethylation and shortened telomeres were found to be significantly associated with early-stage CLL tumors (Binet A) from untreated patients.

A body of work using two mouse models has recently provided strong evidence that aberrant hypo-methylation promotes CLL development. Thus, aberrant hypo-methylation of a single promoter may upregulate several micro RNAs, possibly contributing to tumorigenesis. The TET2 enzyme, conversely, is an enzyme that plays a central role in DNA demethylation by catalyzing the conversion of 5-mC to 5-hydroxymethyl cytosine (5-hmC). Recent studies have suggested strong crosstalk between histone modifications, transcriptional activity, and prior DNA methylation status on DNMT localization.

Treatments with specific methylation agents are used in combination with conventional anti-neoplastic chemo-therapy treatment, Somatic gene therapy has become a research table in clinical trials using a therapeutic DNA, for the treatment of diseases. Recent studies have included the method for the treatment of CLL.

In experimental models, disrupting the MDM2–p53 interaction restored p53 function and sensitized tumors to chemotherapy or radiotherapy. For example, in hematologic malignancies, such as multiple myeloma or chronic lymphocytic leukemia (CLL), the induction of p53 – using a small MDM2-inhibitor molecule, Nutlin-3 – can induce the apoptosis of malignant cells.

Nutlins are a group of cis-imidazoline analogues which have high binding potency and selectivity for MDM2. Nutlin-3 displaces p53 by competing for MDM2 binding. It has also been found that Nutlin-3 potently induces apoptosis in cell lines derived from hematologic malignancies and B-cell CLL with frequent translocation from 14q32 to 17p, with good therapeutic responses.

1.1. Immune therapeutic success

After the chemotherapy treatment, the tumor antigens are taken up by antigen-presenting cells (APCs) and presented in the context of co-stimulatory molecules B7 from malignant cells. T cells recognize antigens to become activated. T cells can differentiate into memory T cells that can be reactivated in the presence of the recurrent tumor. Another way of inhibiting cells identified after the CTLA-4 receptor is represented by PD-1 (programmed to the pH) and PD-L1 or ligands. Like CTLA-4, PD-1 is expressed only in activated T cells to stop their proliferation, limiting the production of the memory T lymphocyte type. However, in contrast to CTLA-4, PD-1 inhibits T cell responses by interfering with T cell receptor signaling, as opposed to out-competing CD28.

There are ongoing clinical trials with anti-CTLA-4 (Ipilimumab, Med-Immune/Astra-Zeneca) plus anti-PD-1 or anti-PD-L1 in other types of tumors with preliminary data showing promising results.

The studies of genotypic and phenotypic chromosome aberrations by the FISH method allow the identification of differential diagnoses in patients with CCL.

The protein ZAP70 is a member of the protein-tyrosine kinase family. ZAP70 is normally expressed in T cells and natural killer cells and has a critical role in the initiation of T-cell signaling. ZAP70 in B cells is used as a prognostic marker in identifying different forms of chronic lymphocytic leukemia (CLL). Some studies have shown that patients with cancer make antibodies against p53 proteins, but the frequency and magnitude of this response are still under debate (Van der Burg et al., 2001). However, a large number of patients with cancer did produce p53-reactive T cells.

A vaccine is delivered in the adjuvant setting and induces T helper-type cells. However, the response may not be potent enough to result in any clinical benefit as a mono-therapy: most patients had T-helper cells that fail to produce key cytokines.

This indicated that these p53-specific T-helper responses are not polarized. Therefore, approaches are being investigated to promote a stronger and more correctly polarized response using both DNA-based and dendritic cell-delivered p53 vaccines.

Elevated c-Myc protein is correlated with increased mutant p53 mRNA in Burkitt's lymphoma cells. If c-Myc can activate the human p53 gene, then one prediction would be that tumor cells expressing elevated c-Myc may express elevated mutant p53 mRNA (Roy et al., 1994; Vassilev et al., 2004).

It was demonstrated that the human p53 promoter is trans-activated by high c-Myc expression. In examining the relative levels of c-Myc and p53 in human Burkitt's lymphomas and other B-lymphoid lines, the hypothesis was made that there is a correlation between the levels of c-Myc protein and p53 mRNA expression. In particular, cells which express very low levels of c-Myc protein also express low levels of p53 mRNA, while cells that express high levels of c-Myc tend to express high levels of p53 mRNA (Starczynowski et al., 2011).

Micro RNAs (miRNAs) are small non-protein-coding RNAs that regulate gene expression by inhibiting the translation or catalyzing the degradation of target mRNAs. Since the first miRNA, lin-4, was identified in 1993, miRNAs have been shown to play critical roles in the regulation of many biological processes, including cell differentiation, proliferation, and apoptosis, with significant influences on normal and malignant hematopoiesis (Chung et al., 2011).

The miRNAs, which are 18 to 25 nanometers in length, bind to the 30-untranslated region (30-UTR) of the target mRNA and thus can regulate a large number of genes simultaneously (Stephens et al., 2011; Lunt et al., 2011).

Therefore, miRNAs act as master gene regulators, similar to transcription factors, and the two may cooperate and ultimately determine gene expression patterns in the cell (Jones et al., 2009; Hatziapostolou et al., 2013). This is an intriguing observation since miR-34a has been identified as a p53 target with pro-apoptotic functions. Thus, it is interesting to speculate that miR-34a may regulate p53 activity, thereby contributing to leukemic stem cell (LSC) development, quiescence, or resistance to therapy.

Various mechanisms of apoptosis resistance in CLL have been described, such as the over-expression of bcl2 mediated by the depletion of inhibitory miR-15 and miR-16.3.

Accumulating evidence indicates that the signal transduction-dependent changes in both glucose energy metabolism and de novo transcription of the D-type cyclin-cdk4/6 pathway are necessary for

quiescent B-lymphocytes to enter the G1-phase of the cell cycle and grow (Shapiro et al., 2006).

Gene mutations, deletions and translocations can be classified as biomarkers of the individual proteomic and genomic profiles of all types of cancer. The structural and biochemical changes in malignant cells will influence the development of characteristic bio-energetic processes that at their level will condition the expression of cellular oncogenes to the detriment of anti-oncogenes that are expressed, normally in an aerobic metabolism.

Blocked apoptosis from malignant diseases may be due to a high ATP concentration originating from anaerobic metabolism initiating the suppression of anti-oncogene proteins, especially p53 protein. Further studies are necessary to detect in patients with high concentrations of ATP the mutations, translocations or deletions of the p53 gene that is located on chromosome 17, using FISH technology.

Phi-Kan083 and PK7088 are small chemicals that bind p53 protein and generate the Y220C mutant, stabilizing it and boosting the amount of wild-type p53 protein. Other molecules, including PRIMA-1, PRIMA met/APR-246, CP-31398, and SCH29074, bind to mutant p53 proteins and interact with DNA-binding domains to facilitate proper mutant protein folding and p53 function recovery. The zinc-binding domain is required for the proper folding of wild-type p53, whereas zinc binding is absent in the R175H p53 mutant.

When zinc is added to the structural mutants G245C and G245D, the wild-type structure is for the most part restored. As a consequence, the discovery of the ability of zinc to restore wild-type folding suggests that this technique might be able to restore anticancer drug chemosensitivity in cells harboring mutant p53 proteins. NSC31926, a thiourea metal chelator, can restore wild-type p53 function in p53 mutant cell lines, most likely via boosting zinc bioavailability to p53 mutants.

Although certain components are designed to selectively inhibit mutant p53, many of them can also interact with and inhibit other members of the p53 protein family, including p63 and p73 proteins.

A small compound called RETRA, which was identified in a screening of a drug used to identify stable wild-type p53, disrupts the interaction between mutant p73 and p53.

Shenzhen SiBiono GeneTech Co. Ltd. produced Gendicine (a recombinant human p53 adenovirus), which was approved by the China Food and Drug Administration (CFDA) in 2003 as a first-in-class gene therapy treatment to treat head and neck cancers. It is a biological therapy that can be administered via intra-tumoral injection, intracavity infusion, or intravascular infusion.

Depending on the cellular stress circumstances, the wild-type p53 protein expressed by Gendicine-transduced cells is a tumor suppressor that is triggered by cellular stress and promotes cell-cycle arrest and DNA repair, or produces apoptosis, senescence, and/or autophagy.

Based on more than 30 published clinical trials and 12 years of commercial use in over 30,000 patients, Gendicine has a proven track record of safety, and when combined with chemotherapy and radiotherapy, it has been shown to produce much higher response rates than traditional therapies alone. In addition to head and neck cancers, Gendicine has been used to successfully treat a variety of other cancer kinds and stages.

Thirteen published trials with long-term survival data found that Gendicine-combination regimens resulted in considerably longer progression-free survival periods than standard treatments alone.

Although many tumors express mutant p53, it is unclear if the many mutations present in this protein have similar activity, and we might have to personalize therapy depending on the presence of a particular mutation rather than only consider whether cancer expresses a wild-type vs a mutant p53.

Previous reports in breast cancer and ovarian cancer cell lines have suggested that PRIMA-1 can reverse mutant p53 aggregate accumulation in cancer cells by substantially decreasing p53 aggregates. Mutant p53-expressing Saos-2 cells also show selective growth-inhibitory and apoptosis-inducing effects when treated with PRIMA-1 (2,2-bis(hydroxymethyl)-1-azabicyclooctan-3-one).

Further, PRIMA-1 and its structural analogue PRIMA-1MET (APR-246) can also inhibit human xenograft tumor growth in SCID mice PRIMA-1 can therefore restore the sequence-specific DNA binding and can activate the functional conformation of mutant p53 in vitro as well as in living cells. PRIMA-1 was observed to rescue both contact and structural p53 mutants. Like PRIMA-1, a maleimide-derived molecule MIRA-1 can also reactivate the DNA binding property of mutant p53 in vitro as well as in cells. In SCID mice, a structural homolog of MIRA-3 had anticancer efficacy against human tumor xenografts expressing mutant p53 **[Scheme 26]**.

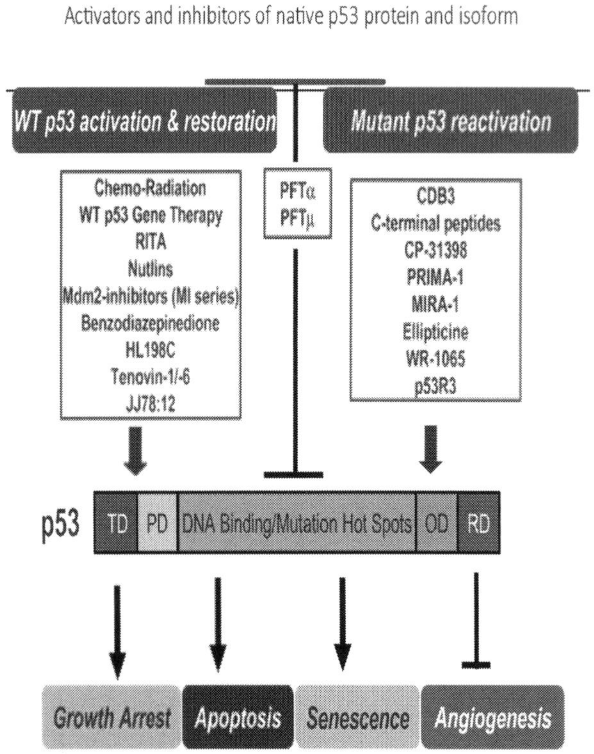

Scheme 26. *Activators and inhibitors of native p53 proteins and isoforms.*

The p53 protein is becoming an attractive therapeutic target for anticancer drug discovery. Three classes of p53-targeting compounds have been identified and characterized. The first class are compounds that activate or restore wild-type p53 function and may be used in human cancers, such as Nutlin-1, a cis-imidazole analogue.

The second class is comprised of PKI-587, a dual competitive ATP inhibitor, which is a therapeutic targeting PI3K (PI3K-α and PI3K-γ) and mTORs. It is an orally bioavailable inhibitor that rescues the p53 protein isomorph carrying a mutation of the p53 gene.

The third class, a heterocyclic compound WYE-354, is a bi-ATP-competitive kinase inhibitor that selectively blocks m-TORC1/2 activity. The growing understanding of mutant p53 actions has contributed to the identification of several compounds with promising therapeutic potential. However, further experiments are required to fully characterize mutant p53 function in cancer. The fact that mutant p53 might play a role in promoting metastasis – the primary cause of cancer-related mortality – is particularly attractive in terms of possible therapies [50].

1.11. Reactivation of p53 protein by microRNAs

miRNAs, discovered in 1993, are generally 19–25 bases in length and do not code for proteins. miRNAs are either overexpressed and act as oncogenes in cancer tissues or they are under-expressed and can suppress proliferation by functioning as tumor suppressors. Nucleic acid miRNA regulation can affect the expression of both wild-type and mutant p53 since those targeting p53 are not able to distinguish between the wild-type and mutant forms if the target site is not the mutation site. Unless a miRNA specifically targets a mutated region of p53 mRNA, it is unable to distinguish between the wild-type and mutant versions of the mRNA.

Some of the miRNAs can regulate the stability of the p53 protein by targeting negative regulators, such as an E3 ubiquitin ligase, MDM2 or an MDM2-related protein MDM4. This results in p53's escape from degradation and its stability in the cell. Reports have shown that miR-192/194/215, miR-143/145, miR-29b, miR-32, miR-605, miR-25, miR-32, miR18b, and miR-339-5p were reported to repress MDM2 and activate p53. However, whether miRNAs can directly bind to amyloid or mutant p53 and activate its function is not known.

Because wild-type p53 is an efficient promoter of apoptosis and senescence tumor cells, reactivating the wild-type characteristics of p53 mutants, which are commonly overexpressed in cancer, is a viable therapeutic strategy.

Several studies have shown that transfection of cancer cells with wild-type p53-expressing plasmids can induce apoptosis and/or growth arrest, implying that a gene therapy method for cancer treatment could be based on restoring normal p53 expression and function.

Several clinical research investigations using viral and non-viral vectors delivering p53 genes, alone or combined with other therapeutic agents, have been completed so far. Some tumor-derived mutations that cause wild-type p53 loss-of-function can be restored by other point mutations that help stabilize the p53 protein, indicating that the structural change is reversible [51].

Summary

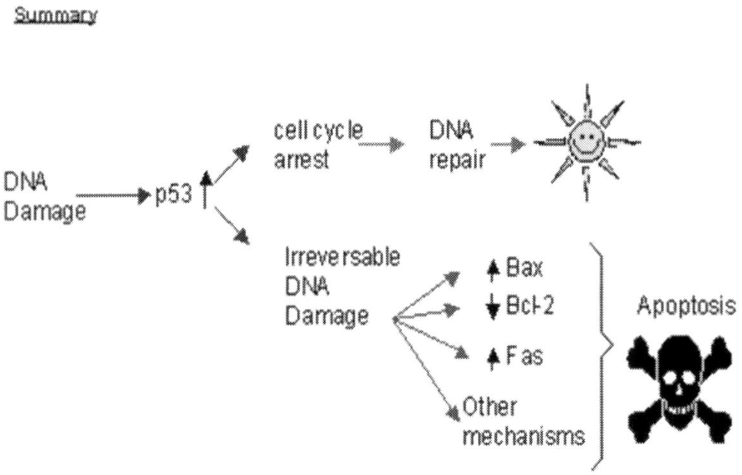

p53 gene mutations are the most common genetic abnormalities of cancer. In the last years, have been extensively studied in various mature B cell malignancies, including CLL to Diffuse Large Lymphoma. Identifying different p53 gene mutations is very important because these mutations have an impact on the patient's clinical course in CLL with the p53 protein mutant isoform.

Figures, Schemes, Photos, Tables and Graphics Legends

Chapter 2

Chapter 3

Chapter 4

Chapter 5

Chapter 6

Chapter 8

Chapter 9

Chapter 10

Abbreviations

CDK -	cyclin-dependent kinase
COX2 -	cyclooxygenase 2
CAF -	fibroblasts associated with colon cancer;
HIF1α -	hypoxia-inducible factor 1α
15-PGDH -	15-hydroxyprostaglandin dehydrogenase,
IL-8 -	interleukin-8
LFS -	Li Fraumen Syndrome
MMP9 -	matrix metalloproteinase 9
MAPKP5 -	mitogen-activated protein kinase phosphatase-5
NF-κB -	nuclear factor-κB
PG -	prostaglandin
PTEN -	phosphatase and tensin homolog;
TERT -	reverse telomerase transcriptase
TIMP1 -	tissue inhibitor of metalloproteinases
1-TGFβ -	transforming growth factor
VEGF -	vascular endothelial growth factor

References

1. Katharine E Brock. Li-Fraumeni Syndrome. Medscape Reference. October 2013; http://emedicine.medscape.com/article/987356-overview. Accessed in 02-02-2023.
2. Kern SE, Kinzler KW, Bruskin A, Jarosz D, Friedman P, Prives C, Vogelstein B. Identification of p53 as a sequence-specific DNA-binding protein. Science. 1991; (5013): 1708–11.
3. Hiwase D, Hahn C, Ngoc E, Tran H, Chhetri R, Baranwal A, Al-Kali A. TP53mutation in therapy-related myeloid neoplasm defines a distinct molecular subtype. Blood. 2023; 141 (9); 2023: 1087-91.
4. May PE. Twenty years of p53 research: structural and functional aspects of the p53 protein. Oncogene. 1999; 18 (53): 7621–36.
5. Kim ST, Lim DS, Canman CE, Kastan MB. Substrate specificities and identification of putative substrates of ATM kinase family members. The Journal of Biological Chemistry. 1999; 274 (53): 37538–43.
6. Peng Y, Chen L, Li C, Lu W, Chen J. Inhibition of MDM2 by hsp90 contributes to mutant p53 stabilization. The Journal of Biological Chemistry. 2001; 276(44): 5635-46.
7. Udristioiu A, Florescu C, Popescu MA, Cojocaru M. High Concentration of anaerobic ATP implicated in aborted apoptosis from CLL. LabMed. 2010; 41: 203-8.
8. Cabrita MA, Bos eR, Vanzyl EJ, Alyssa P, Marcellus KA, Pan E. The p53 protein induces stable miRNAs that have the potential to modify. subsequent p53 responses. Gene. 2017; 608: 86–94.
9. Hallek M. Therapy of chronic lymphocytic leukaemia. Best Pract Res Clin Haematol. 2010;23: 85–96.
10. Mraz M, Malinova K, Kotaskova J, Pavlova S, Tichy B, Malcikova J, Stano Kozubik K, Smardova J, Brychtova Y. MiR-34a, miR-29c and miR-17-5p are downregulated in CLL patients with TP53 abnormalities. Leukemia: official journal of the Leukemia Society of America, Leukemia Research Fund, UK 2009; 23: 1159–63.
11. Tang J, Di J, Cao H, Bai J, Zheng J. p53-mediated autophagic regulation: A prospective strategy for cancer therapy. Cancer Lett. 2015; 363(2):101-7.
12. E Tasdemir, MC Maiuri, L. Galluzzi, I. Vitale, M. Djavaheri-Mergny et al. Regulation of autophagy by cytoplasmic p53. Nat. Cell Biol. 2008; 10; 676-87.

13. O'Farrell F, Rusten TE, Stenmark H. Phosphoinositide 3-kinases as accelerators and brakes of autophagy. FEBS J 2013; 280(24):6322–37.
14. Levine B, Sinha S, Kroemer G. Bcl-2 family members: dual regulators of apoptosis and autophagy. Autophagy. 2008; 4(5):600–6.
15. Yin Z, Pascual C, Klionsky DJ. Autophagy: machinery and regulation. Microb Cell. 2016 3(12):588–96.
16. Galluzzi L, Pietrocola F, Bravo-San Pedro JM, Amaravadi RK, Baehrecke EH, Cecconi F, et al. Autophagy in malignant transformation and cancer progression. EMBO J 2015; 34(7):856–80.
17. X.H. Liang, S. Jackson, M. Seaman, K. Brown, B. Kempkes, H. Hibshoosh, B. Levine. Induction of autophagy and inhibition of tumorigenesis by Beclin 1. Nature. 1999; 402; 672-6.
18. Arsov I, Adebayo A, Kucerova-Levisohn M, Haye J, MacNeil M, Papavasiliou FN, et al. A role for autophagic protein beclin1 early in lymphocyte development. J Immunol. 2011;186(4):2201–9.
19. Z. Feng, W. Hu, E. de Stanchina, A.K. Teresky, S. Jin, S. Lowe, A.J. Levine. The regulation of AMPK beta1, TSC2, and PTEN expression by p53: stress, cell and tissue specificity, and the role of these gene products in modulating the IGF-1-AKT-mTOR pathways. Cancer Res. 2007;67: 3043-53.
20. S. Arico, A. Petiot, C. Bauvy, P.F. Dubbelhuis, A.J. Meijer, P. Codogno, E. Ogier-Denis. The tumor suppressor PTEN positively regulates macroautophagy by inhibiting the phosphatidylinositol 3-kinase/protein kinase B pathway. J. Biol. Chem 2001; 276: 35243-6.
21. Levine B, Klionsky DJ. Development by self-digestion: molecular mechanisms and biological functions of autophagy. 2004; 6(4):463–77.
22. Klionsky DJ, Abdelmohsen K, Abe A, Abedin MJ, Abeliovich H, Acevedo Arozena A, et al. Guidelines for the use and interpretation of assays for monitoring autophagy (3rd edition). Autophagy 2016; 12(1):1–222.
23. Nelson BP, Gupta R, Dewald GW, Paternoster SF; Rosen ST, Peterson LC. Chronic Lymphocytic Leukemia FISH Panel: Impact on Diagnosis. Am J Clin Pathol. 2007;128(2):323-32.
24. Dhanasekaran DN, Reddy EP. JNK Signaling in Apoptosis. Oncogene. 2008; 20; 27(48): 6245–51.
25. Pomo MJ, Taylor MR, Gullap RR. A model system to understand cancer cell growth mechanics. Cancer Cell International 2016; 16:1-44.
26. Bennett JM. Practical of Diagnosis of Hematologic Disorders. Lett. Molecular Prognostic Factors in Non – Hodgkin Lymphoma, Chronic Lymphocytic Leukemia, Chicago, ASCP Press. 2006; p.615 – 71.

27. Pomo MJ, Taylor MR, Gullap RR. A model system to understand cancer cell growth mechanics. Cancer Cell International 2016; 16:1-44.
28. Sabapathy K. The Contrived Mutant p53 Oncogene – Beyond Loss of Functions. Sci Rep. 2015; 7: 99913.
29. Nelson BP, Gupta R, Dewald GW, Paternoster SF; Rosen ST, Peterson LC. Chronic Lymphocytic Leukemia FISH Panel: Impact on Diagnosis. Am J Clin Pathol. 2007;128(2):323-32.
30. Whibley C, Pharoah P, Hollstein M. P53 Polymorphism: Cancer Implications. Nat Rev Cancer. 2009; (2): 95-107.
31. Glassford J1, Soeiro I, Skarell SM, Banerji L, Holman M,. BCR targets cyclin D2 via Btk and the p85alpha subunit of PI3-K to induce cell cycle progression in primary mouse B cells. Oncogene. 2003; 17;22.
32. Luijsterburg, M. S.; van Attikum, H. Chromatin and the DNA damage response: The cancer connection. Mol. Oncol., 2011, 5(4): 349-67.
33. Yamada O, Kawauchi K. The role of the JAK-STAT pathway and related signal cascades in telomerase activation during the development of hemato- malignancies. JAKSTAT. 2013; 2(4): e25256.
34. Seif F, Khoshmirsafa M, Aazami H, Mohsenzadegan M, Sedighi M. The role of JAK-STAT signaling pathway and its regulators in the fate of T helper cells. Cell Commun Signal. 2017; 15: 23.
35. Ross SH, Cantrell DA. Signaling and Function of Interleukin-2 in T Lymphocytes. Annu Rev Immunol. 2018; 26 (36): 411–33.
36. Ross SH, Cantrell DA. Signaling and Function of Interleukin-2 in T Lymphocytes. .Ann Rev Immunol. 2018; 26 (36): 411–33.
37. Bencik, J, Thanh TH, Johannes P, Javaheri T, Schlederer M. JAK-STAT signaling in cancer: From cytokines to non-coding genome. Cytokine. 2016.
38 Yang I, Nie J, Ma X, Wei, Y Peng Y, Wei X.]. Targeting PI3K in cancer: mechanisms and advances in clinical trials. Molecular Cancer 2019; 26.
39. Rane CK, Minden A. P21 activated kinases, Structure, regulation, and functions. Small GTPases. 2014; 5: e28003.
40. Klotza,n LO, Ramosb CS, Arroyob IP,Holger PU, Redox Biology. Redox regulation of Fox. O transcription factor. 2015; Available online 3 July 2015.
41. Maria-Magdalena Georgescu. PTEN Tumor Suppressor Network in PI3K-Akt Pathway Control. Genes Cancer. 2010; 1(12): 1170–77.
42. Michael Hallek,1,2 Bruce D. Cheson,3 Daniel Catovsky,4 Federico Caligaris-Cappio,5 Guillermo Dighiero. CLL, guidelines for diagnosis, indications for treatment, response assessment, and supportive management of CLL. Blood 2018; (131): 25-4.

43. Abdel-Magid AF. Reactivation of the guardian of the genome P53: a promising strategy for treatment of cancer. ACS Med Chem Lett. 2021;12 :331–3.
44. Batır MB, Şahin E, Çam FS. Evaluation of the CRISPR/Cas9 directed mutant TP53 gene repairing effect in human prostate cancer cell line PC-3. Mol Biol Rep 2019;46: 6471–84.
45. Enache OM, Rendo V, Abdusamad M, Lam D, Davison D, Pal S, et al. Cas9 activates the p53 pathway and selects for p53-inactivating mutations. Nat Genet 2020; 52:662–8.
46. Lynette A. Smyth⊠ and Ian Collins. J Chem Biol. 2009 Aug; 2(3): 131–51.
47. Udristioiu. A. Clinical Trail:ISRCTN12539707: https://doi.org/10.1186/ISRCTN12539707 .Clinical Trail: The role of the P53 gene and the p53 protein in non-Hodgkin malignant lymphomas. Registration date 12/11/2022.
48. Akritopoulou-Zanze I, Hajduk PJ. Kinase-targeted libraries: The design and synthesis of novel, potent, and selective kinase inhibitors. Drug Discov Today 2009;14(5–6):291–297.
49. Andersen CB, Wan Y, Chang JW, Riggs B, Lee C, Liu Y, Sessa F, Villa F, Kwiatkowski N, Suzuki M, Nallan L, Heald R, Musacchio A, Gray NS. Discovery of selective aminothiazole Aurora kinase inhibitors. ACS Chem Biol. 2008;3(3):180–92.
50. Apsel B, Blair JA, Gonzalez B, Nazif TM, Feldman ME, Aizenstein B, Hoffman R, Willi Kokat KM, Knight ZA. Targeted polypharmacology: discovery of dual inhibitors of tyrosine and phosphoinositide kinases. Nat Chem Biol. 2008;4(11):691–99.
51. Marei HE, Althani A, Afifi N, Hasan A, Caceci T, Pozzoli G. p53 signaling in cancer progression and therapy. Cancer Cell International. 2021; 21:703.

Made in United States
Orlando, FL
04 November 2026

90207776R00092